FUNDAMENTAL THEORIES IN PHYSICS

Studies in the Natural Sciences

A Series from the Center for Theoretical Studies

University of Miami, Coral Gables, Florida

ORBIS SCIENTIAE

FUNDAMENTAL THEORIES IN PHYSICS

Orbis Scientiae Moderators
Donald Glaser
Behram Kursunoglu
Sidney Meshkov
Edward Teller

Editors
Stephan L. Mintz
Laurence Mittag
Susan M. Widmayer

Scientific Secretaries
Chui-Shuen Hui
Joseph F. Malerba

Center for Theoretical Studies
University of Miami
Coral Gables, Florida

PLENUM PRESS • NEW YORK AND LONDON

Library of Congress Cataloging in Publication Data

Orbis Scientiae, University of Miami, 1974.
Fundamental theories in physics.

(Studies in the natural sciences, v. 5)
Sponsored by Center for Theoretical Studies, University of Miami, Coral Gables, Fla.
1. Nuclear reactions — Congresses. 2. Particles (Nuclear physics) — Congresses. 3. Space and time — Congresses. 4. Stars — Evolution — Congresses. I. Mintz, Stephan, ed. II. Mittag, Laurence, ed. III. Widmayer, Susan M., ed. IV. Miami, University of, Coral Gables, Fla. Center for Theoretical Studies. V. Title. VI. Series.
QC793.9.072 1974 530 74-9659
ISBN 0-306-36905-2

Part of the Proceedings of Orbis Scientiae held by the Center for Theoretical Studies, University of Miami, January 7-12, 1974

A Division of Plenum Publishing Corporation
227 West 17th Street, New York, N.Y. 10011

United Kingdom edition published by Plenum Press, London
A Division of Plenum Publishing Company, Ltd.
4a Lower John Street, London, W1R 3PD, England

Printed in the United States of America

Preface

This volume includes the papers presented during the high energy session of the first Orbis Scientiae held by the Center for Theoretical Studies, University of Miami. The Orbis met from January 7th to 12th, 1974 for the purpose of allowing scientists in various disciplines to acquaint each other with frontier problems of their respective disciplines. This Orbis Scientiae was the first of a new series of annual gatherings which have replaced the Coral Gables Conferences. These new meetings will include, in addition to physics, the mainstay of the Coral Gables Conferences, other sciences and will allow scientists in the various disciplines represented to exchange views with each other as well as among themselves.

The high energy sessions of the Orbis were devoted primarily to fundamental theories which attempt to unify a number of interactions in physics such as the electromagnetic and gravitational forces, the electromagnetic and weak forces, or indeed all of the known interactions. Papers were also invited on other topics of current interest in high energy physics and astrophysics and they form part of this volume.

This volume is not subdivided but the first half is devoted to fundamental theories and the second half to other subjects of current interest. The last paper, on stellar evolution, was presented at the Sixth Annual J. Robert Oppenheimer Memorial Prize Ceremonies by the recipient, Professor Edwin E. Salpeter of Cornell University.

The editors would like to express their appreciation to Mrs. Helga Billings, Miss Sara Lesser and Mrs. Jacquelyn Zagursky for their industrious typing of the manuscripts and assistance with the details of the meeting.

The Orbis Scientiae was supported in part by the Atomic Energy Commission.

Contents

THE GEOMETRICAL NATURE OF SPACE AND TIME

P. A. M. Dirac

Department of Physics

The Florida State University

Tallahassee, Florida 32306

It used to be considered obvious that space is Euclidean and time is an independent variable. Einstein changed that.

First with special relativity he showed that space and time must be considered together as a 4-dimensional space. It is subject to a somewhat different geometry from Euclid's, called Minkowski geometry, in which there is a minus sign in the formula for the squared length of the hypotenuse of a right-angled triangle.

With general relativity Einstein challenged the Euclidean character even of 3-dimensional space. He supposed space to be curved and to be approximately Euclidean in a small region. The curved space is of a kind that can be immersed in a flat space of a higher number of dimensions. Such a curved space is called a Riemann space, and was extensively studied by mathematicians in the last century.

The complete Einstein theory requires physical space to be a 4-dimensional Riemann space which is approximately Minkowski space in a small region. It led to an explanation of gravitation. This provided the justification for the assumptions.

Physicists then had to reformulate all their laws in such a Riemann space. This proved to be possible, and not too difficult (apart from the quantum laws, for which there are difficulties that have still not been resolved).

It has led to a picture of the physical world in which we deal with particles and fields embedded in a Riemann space. This picture has survived up to the present time -- over fifty years. There is no experimental evidence against it. But still, theoretical physicists are not satisfied with it, and are wondering whether the underlying space should not be more general.

The passage from Euclidean to Riemann space was so successful. Should one not try to take another step in the same direction?

Passing to Riemann space provided one with an excellent description of the gravitational field. Now there is another field very much like the gravitational field, the electromagnetic field. Both fields involve long-range forces, which fall off according to the law r^{-2}. This distinguishes them from the other fields of physics, which are specially important for atomic physics, involving short-range forces, which fall off according to the law $e^{-r/a}$.

Should one not treat the electromagnetic field like the gravitational, and try to explain it in terms of a more general geometry? This leads to the problem of the unification of the gravitational and electromagnetic fields.

Many attempts have been made to solve this problem. The first was Weyl's, which came out very shortly after Einstein's theory of gravitation. I would like to explain the basic idea of Weyl's geometry.

WEYL'S GEOMETRY

The curvature of space required by Einstein's theory can be discussed in terms of the notion of the parallel displacement of a localized vector. The transport of a vector around a closed loop by parallel displacement results in the final direction of the vector being different from its initial direction. Weyl's generalization was to suppose the final vector has a different length as well as a different direction. So long as one is dealing with a Riemann space, the vector can be considered as moving in a flat space of a higher number of dimensions and its length cannot change. Weyl's assumption provided a natural generalization of Riemann space.

With Weyl's geometry there is no absolute way of comparing elements of length at two different points, unless the points are infinitely close together. The comparison can be made only with respect to a path joining the two points, and different paths will lead to different results for the ratio of the two elements of length. In order to have a mathematical theory of lengths one must set up arbitrarily a standard of length at each point, and then refer any length that turns up in the description of space at some point to the local standard at that point. One then as a definite value for the length of a vector at any point, but this value changes when one changes the local standard of length. The standard of length can, of course, be changed by a factor that varies from place to place.

Consider a vector of length ℓ situated at a point with coordinates x^μ (μ=0,1,2,3). Suppose it is transported by parallel displacement to the point $x^\mu + \delta x^\mu$. Its change in length $\delta\ell$ will be proportional to ℓ and to the δx^μ, so it will be of the form

$$\delta\ell = \ell\ \kappa_\mu \delta x^\mu. \tag{1}$$

We have some coefficients κ_μ appearing here. They are further field quantities, occurring in the theory together with the Einstein $g_{\mu\nu}$ and just as fundamental.

Let us suppose the standards of length are changed so that lengths get multiplied by the factor $\lambda(x)$, depending on the x's. Then ℓ gets changed to $\ell' = \ell\lambda(x)$ and $\ell + \delta\ell$ gets changed to

$$\ell' + \delta\ell' = (\ell+\delta\ell)\ \lambda(x+\delta x)$$

$$= (\ell+\delta\ell)\ \lambda(x) + \ell(\partial\lambda/\partial x^{\mu})\ \delta x^{\mu}.$$

with neglect of second order quantities. We now find

$$\delta\ell' = \ell'\kappa'_{\mu}\delta x^{\mu}$$

with

$$\kappa'_{\mu} = \kappa_{\mu} + \partial \log \lambda/\partial x^{\mu}. \qquad (2)$$

It will be seen that the new field quantities κ_{μ} appearing in Weyl's theory may be taken to be the electromagnetic potentials. They have all the desired properties. They are subject to the transformations (2) arising from a change in the choice of the artificial standards of length. Weyl's geometry provides just what one needs for describing both the gravitational and electromagnetic fields in geometrical terms.

In spite of these beautiful features of the theory, it was not acceptable to physicists, because it clashes with the quantum theory. Quantum phenomena provide an absolute standard of length. An atomic clock measures time in an absolute way, and if one takes the velocity of light to be unity one gets an absolute standard of distance. There is thus no need for the arbitrary metric standards of Weyl's theory. One may suppose that under parallel displacement a vector keeps the same length with respect to these atomic standards and then the variation $\delta \ell$ of equation (1) does not arise.

With the rejection of Weyl's theory physicists returned to the view that the gravitational field influences the geometry of space, but the electromagnetic field is only something immersed in the space established by the gravitational field. This view provided a satisfactory working basis. But many physicists remained fascinated by the problem of finding a geometrical interpretation of the electromagnetic field and so obtaining a unified field theory. Many theories on these lines have been proposed, but they are all complicated and rather artificial, and are not generally accepted. Weyl's theory remains as the outstanding one, unrivaled by its simplicity and beauty. Some recent developments lead to a possibility that it may be revived.

THE LARGE NUMBERS HYPOTHESIS

From the constants of nature one can construct some dimensionless numbers. The important ones are $\hbar c/e^2$, which is about 137, and the ratio of the mass of the proton to that of the electron,

M/m, which is about 1840. There is no explanation for these numbers, but physicists believe that with increasing knowledge an explanation will some day be found.

Another dimensionless number is provided by the ratio of the electric to the gravitational force between an electron and a proton, namely e^2/GMm. This has a value about 2×10^{39}, quite a different order from the previous ones. One wonders how it could ever be explained.

The recession of the spiral nebulae provides an age for the universe of about 2×10^{10} years. If one expresses it in terms of some atomic unit, say e^2/mc^3, one gets a number about 7×10^{39}, which is comparable with the previous large number. It is hard to believe that this is just a coincidence. One suspects that there is some connection between the two numbers, which will get explained when we have more knowledge of cosmology and of atomic theory.

One can set up a general hypothesis, which we may call "the Large Numbers Hypothesis", that all dimensionless numbers of this order that turn up in nature are connected. One of these large numbers is the epoch t, the present time reckoned from the time of creation as zero, and this increases with the passage of time. The Large Numbers Hypothesis now requires that they shall all increase, in proportion to the epoch, so as to maintain the connection between them. One can infer that the gravitational constant G, measured in atomic units, must be decreasing in proportion to t^{-1}.

Now Einstein's theory of gravitation requires that G shall be constant; in fact with a suitable choice of units it is 1. Thus Einstein's theory of gravitation is irreconcilable with the Large Numbers Hypothesis.

The Large Numbers Hypothesis is a speculation, not an established fact. It can become established only by direct observation of the variation of G. The effect is not too small to be beyond the capabilities of present-day techniques. Shapiro's (Scientific American, 219, 32, 1968) measurements of the distances of the planets by radar are extremely accurate, and if G is really varying to the required extent, it should show up in his observations in a few years time.

Also, the distance of the moon can now be observed with very great accuracy by the reflection of laser beams. This distance is subject to many disturbances, but these can be estimated and allowed for. If there is some residual effect which cannot be otherwise explained, it would have to be ascribed to a variation of G.

There is also the possibility of observing the variation of G directly by laboratory experiments. Dr. Beams is engaged in the accurate measurement of G, and he considers that it may be possible to improve his apparatus, immersing it in a bath of liquid helium, so that a variation of G of the expected order would show up (Physics

Today, May 1971, p. 35).

We shall assume that the Large Numbers Hypothesis is correct so that G does vary, and consider how Einstein's theory can be modified to agree with it.

THE TWO METRICS

A simple way of effecting a reconciliation is to suppose that the Einstein equations refer to an interval ds_E connecting two neighboring points which is not the same as the interval ds_A measured by atomic apparatus. By taking the ratio of ds_E to ds_A to vary with the epoch we get G varying with the epoch. The ratio is sufficiently nearly constant for the modification in Einstein's theory to be very small.

With the introduction of the two metrics ds_E and ds_A, we see that the objection to Weyl's theory discussed previously falls away. One can apply Weyl's geometry to ds_E, supposing that it is non-integrable when transported by parallel displacement, so that we must refer it to an arbitrary metric gauge to get a definite value for it. Then ds_E gets altered when we make a transformation (2) of the potentials. On the other hand ds_A is referred to atomic units and does not depend on an arbitrary metric gauge and is not affected by a transformation (2) of the potentials.

The measurements ordinarily made by physicists

in the laboratory use apparatus which is fixed by the atomic properties of matter, so the measurements will refer to the metric ds_A. The metric ds_E cannot be measured directly, but it shows itself up through its influence on the equations of motion. It forms the basis of all dynamical theory, whether the theory is the accurate one of Einstein or the Newtonian approximation. The relation of the two metrics is exemplified by radar observations of the planets. Here a distance which is determined by equations of motion is measured by atomic apparatus.

Let us consider the relation between the two ds's. We must refer again to the Large Numbers Hypothesis. We bring in another large dimensionless number, the total number of nucleons in the universe. If the universe is infinite, we replace this by the total number of nucleons in the part of the universe that is receding from us with a velocity less than $\frac{1}{2}c$.

This number is rather uncertain because we do not know how much invisible matter there is. Presumably the amount of invisible matter is not very much larger than the visible. We then get a number somewhat around 10^{78}. From the Large Numbers Hypothesis, this must vary in proportion to t^2. It follows that new protons and neutrons are continually being created.

The question arises, Where are they created? There are two reasonable assumptions one might make.

1. Matter is created uniformly throughout space, hence mostly in intergalactic space. We may call this <u>additive creation.</u>

2. Matter is created where it already exists, in proportion to the amount existing there. We may call this <u>multiplicative creation.</u>

I do not know which assumption to prefer. In any case the creation of matter is a new physical process, not explicable in terms of any of the known physical processes. The effect is too small to be easily detectable, except possibly in case 2. This assumption would require all matter to multiply, including the matter in the earth. There might be some difficulty in understanding how the matter in very old rocks can have multiplied without disrupting their crystal structure.

Let us return to the relation between ds_E and ds_A. Take as an example the motion of the earth around the sun. According to Newton, the basic formula is

$$GM = v^2 r$$

where M is the mass of the sun, r is the radius of the earth's orbit and v is the velocity. The formula must hold both in Einstein and atomic units. In Einstein units all the quantities are constants. In atomic units they may vary with

t. Let us write the formula then

$$G_A M_A = v_A^2 r_A.$$

The velocity v_A is a certain fraction of the velocity of light, the same fraction as in Einstein units, and is thus constant. We have already seen that G_A is proportional to t^{-1}. With additive creation M_A is constant, and so r_A is proportional to t^{-1}. With multiplicative creation M_A is proportional to t^2 and r_A is proportional to t.

Thus with additive creation the earth is approaching the sun, and the whole solar system is contracting. With multiplicative creation the earth is receding and the whole solar system is expanding. These effects are cosmological, and are to be superposed on other effects arising from known physical causes. Shapiro's observations or accurate lunar observations should show them up.

We get finally for the relation between the two metrics

$$ds_E = t\,ds_A \quad \text{additive creation,}$$

$$ds_E = t^{-1} ds_A \quad \text{multiplicative creation.}$$

Of course with Weyl's geometry these relations are only approximate, because ds_E is affected by changes in the metric gauge accompanied by changes in the electromagnetic potentials, while ds_A is unaltered by such changes.

SYMMETRY BREAKING

The Weyl interpretation of the electromagnetic field as influencing the geometry of space and not merely as something immersed in Riemannian space has a striking consequence -- symmetry breaking. Consider a charged particle and take a field point P close to its world-line. For simplicity, suppose the coordinate system to be chosen so that the particle is momentarily at rest.

Now take an element of length ℓ at P and suppose it to be shifted by parallel displacement into the future, by an amount δx^o. From the fundamental formula (1), it will change by

$$\delta\ell = \ell\, \kappa_o \delta x^o .$$

Here κ_o will consist mainly of the Coulomb potential arising from the charged particle. Suppose the sign of the charge is such that ℓ increases when it is shifted into the future. With the opposite sign of the charge it will decrease. Now there is no symmetry between a quantity increasing and the same quantity decreasing. Consequently there is no symmetry between positive and negative charge.

If ℓ increases when it is shifted into the future, it decreases when it is shifted into the past. So there is no symmetry between future and past. But if one changes the sign of the charge and also interchanges future and past, one gets back to the original situation.

Atomic physicists have introduced the operators P for changing the parity, C for charge conjugation and T for time reversal. In elementary theories all these symmetries are preserved. The present theory does not provide any breaking of the P symmetry. However, it does break the C and T symmetries, while preserving their product CT.

Experimentally, all three symmetries are observed to be broken, but the product PCT is conserved, so far as is known. It would seem that the breaking of the P symmetry must be ascribed to the short range atomic forces. However, the breaking of C and of T, with preservation of CT, is caused by the long range forces, if they are handled in accordance with Weyl's geometry. It would seem that this symmetry breaking arises from the interaction of the gravitational and electromagnetic fields. The effect must be small, because gravitational effects are always small in the atomic domain.

COSMOLOGICAL MODELS

The Large Numbers Hypothesis puts severe restrictions on the permissible models of the Universe. A favorite model is to suppose that the Universe started off as something very small, possibly a point, that it increases up to a certain maximum size and then collapses again. Many people believe in that model. But it is quite unacceptable if one believes in the Large Numbers Hypothesis. With that model there is a maximum size to the Universe. Take this maximum size and express it in terms of some atomic unit, and you get a large number which is a constant. It is just such a constant large number which is ruled out by the Hypothesis.

Thus we must have the Universe expanding and continuing to expand forever. Furthermore, there cannot be any change in the law of expansion, except at the very early stages, because such a change would involve a time when the change becomes effective and this time, expressed in atomic units, would again give us a large constant number. We are left with only very simple models.

One can work out what models are permissible, with either of the assumptions for the continuous creation of matter. I will just give the results of the calculations. The arguments are all quite simple.

One finds that, if one assumes multiplicative creation, one is forced to a picture which, referred to the Einstein units, for which the Einstein equations hold, is just Einstein's cylindrical model of the Universe. The Universe is then closed with a finite constant size referred to ds_E.

The apparent recession of the nebulae in this model arises from the fact that, when we go over to an atomic standard of measurement, atomic clocks are continually speeding up with reference to the Einstein unit of time. Thus, if we take a distant stellar object, emitting light to us of a certain spectral frequency, the wavelength refers to the atomic unit of length at the time when it was emitted. That wavelength remains constant, in Einstein units, as the light comes towards us. When it arrives here it is compared with the present atomic clocks, which are going faster, so the wavelength appears to have increased. That is the explanation for the apparent recession of the spiral nebulae, on a model based on Einstein's steady Universe. That is what we have with multiplicative creation.

What do we have with additive creation? We are now forced to an entirely different picture. You see, if we work in terms of the Einstein units we must have the Einstein equations holding and the Einstein equations demand conservation of mass. How is that to be fitted in with the continuous creation of protons and neutrons? With multiplicative creation we use the device of keeping the Einstein mass of a body the same, through having the nucleon mass getting smaller, referred to the Einstein unit of mass, according to a law that just compensates for the increasing number of nucleons. The mass of a body such as the sun, referred to Einstein units, then remains constant.

Now with additive creation, we cannot preserve conservation of mass by any choice of our unit of mass. We have the new matter continually being created in interstellar space. The only way I can think of for preserving the conservation of mass, which is demanded by the Einstein theory, is to suppose that simultaneously with the creation of nucleons there is a creation of some kind of negative mass which is unobservable, and is such that the total creation of mass is zero and fits in with the Einstein theory. We thus have two kinds of matter being created. The matter which we can observe in the form of protons and neutrons, quantized matter one might call it; and the negative mass which is unobservable, which is uniformly distributed over the whole of space, and which has no physical effects, except for the gravitational effects, causing a curvature of space.

You might say that this picture is very artificial and unlikely, but it seems that we are really forced to it from the one assumption of the Large Numbers Hypothesis, together with our requiring that the Einstein theory shall hold with respect to a suitable ds_E. It is not an assumption which is introduced arbitrarily at a later stage in the theory. It follows directly from the early assumptions. It requires two kinds of matter to be created, a continuously spread out, negative mass, which compensates for the quantized mass in the form of protons and neutrons which we can observe.

The net effect of this continuous creation is that when we smooth out local irregularities space-time is flat. We have then, roughly, Minkowski space. In this Minkowski space there is somewhere a point where the origin of the Universe occurred, where the big bang occurred, which started off the physical world. All the physical world that we can observe lies in the future light cone of this point of origin. All the galaxies radiate out within the light cone, from the point of origin, and also the uniform background of negative mass, which results in the total density being approximately zero. It's really a very simple model, much simpler than what cosmologists usually work with.

So far as I can see we are forced to one of these two alternatives, the model based on Minkowski space and the Einstein cylindrical model, corresponding to the two kinds of creation of matter.

UNIFIED THEORY OF FUNDAMENTAL INTERACTIONS*,**

Behram Kursunoglu

Center for Theoretical Studies

University of Miami, Coral Gables, Fla.

SUMMARY OF RESULTS

1. All elementary particles carry a magnetic charge g (different magnitudes for different particles) associated with a short range field. This charge is superimposed over a magnetically neutral particle core consisting of a distribution of magnetic charge density in stratified layers of sharply decreasing magnitudes and alternating signs. Magnetic monopoles associated with a long range field are shown not to be present.

---

*A shorter version of this paper entitled "Gravitation and Magnetic Charge" will appear in the Physical Review D May 15, 1974 issue.

**Presented in the January 1974 Orbis Scientiae of the Center for Theoretical Studies, University of Miami, Coral Gables, Florida 33124.

2. The magnetic charge g assumes an infinite spectrum of values and is an invertible function of mass. For $g \neq 0$ the fields are regular everywhere. For $g = 0$ the solutions collapse into the solutions of general relativity and classical electrodynamics.
3. The self-energy of the elementary particles are finite.
4. As a consequence of the general covariance of the theory the surfaces of zero magnetic charge density in the particle core have an indeterminacy extending over a nucleon Compton wave length. These relationships lead to a mass spectrum.
5. For every positive energy solution there exists also a negative energy solution with corresponding electric and magnetic charges.
6. The presence of negative energy solutions along with positive energy solutions imbedded in a gravitational field point to a large scale symmetry with respect to a distribution of matter and anti-matter in the universe.

UNIFIED THEORY OF FUNDAMENTAL INTERACTIONS

Behram Kursunoglu

Center for Theoretical Studies

University of Miami, Coral Gables, Fla.

ABSTRACT

The field equations of the generalized theory of gravitation which were proposed over 20 years ago by this author have now been solved for the static spherically symmetric case. It is found that electric and magnetic charges are two fundamental constants of integration and that the corresponding electric, magnetic and gravitational fields are regular everywhere only if the magnetic charge $g \neq 0$. The magnetic charge g assumes an infinite spectrum of values and is a function of mass. For $g = 0$, the solutions reduce to the Nordström solution of general relativity in the limit of large r. The theory leads to elementary particles of finite self-energy ($\Delta(\pm E) = mc^2 - \frac{(2g_o)^2}{\ell_o}$) and binding energy. The structure of an elementary particle which is due to the existence of finite $\pm g$ consists of a magnetically neutral core of matter containing a distribution of magnetic charge density in stratified layers of sharply decreasing magnitude and alternating signs so that magnetic monopoles associated with a long range field do not exist. As a consequence of the general covariance of the theory the surfaces of zero magnetic charge density in the core of an elementary particle have an indeterminacy. These facts lead to a mass spectrum for the elementary particles. In addition to charged electric and magnetic currents, the theory yields electrically and magnetically neutral currents and the corresponding fields. The neutral current and the corresponding neutral field are the classical counterparts of the vacuum polarization in quantum electrodynamics. For every positive energy solution there exists also a negative energy solution with the corresponding electric and magnetic charges. For $g = 0$, the volume integral of the neutral electric current density diverges. The assymmetry of Maxwell's equations with regard to the absence of a magnetic current can be understood because the neutral and charged magnetic currents are localized in the core of elementary particle generating short range fields alone.

Furthermore, the theory yields two lengths of the dimensions of 10^{-25} cm and 10^{-15} cm which could serve to differentiate between leptonic and hadronic processes as well as a length 10^{-34} cm referring to the minimum size to which an elementary particle could have collapsed (micro-black-holes).

The presence of negative energy solutions along with positive energy solutions point to a large scale symmetry with respect to a distribution of matter and antimatter in the universe.

I. INTRODUCTION

The success of the renormalization scheme in quantum electrodynamics has brought about a strong feeling of security amongst physicists and it led to the adoption of "renormalizability" of a physical theory as a basic principle of physics. In fact, renormalization has not solved the real problem but it has circumvented the difficulties in quantum electrodynamics. The problem of infinite self-energy integrated into the fiber of classical and quantum electrodynamics defied all attempts for a finite solution of this basic difficulty. The point structure of an elementary particle was a necessary offspring of a mechanical theory where relativistic invariance for a particle with an extended structure was not possible. The theory of renormalization of electric charge and mass has, actually, concealed the fundamental properties (fundamental interactions, mass spectrum, basic symmetries) of the elementary particles arising, essentially from the existence of their extended structures. In this paper I shall give an example to illustrate the interesting consequences for the fundamental interactions of elementary particles associated with the finiteness of the self-energy. It is pleasing to see that a theory without any infinities does, in fact, provide a unification of all fundamental interactions. It will be demonstrated that not only gravitation and electromagnetism but all fundamental interactions have classical counterparts emerging from a single field which is regular everywhere.

Gravitational and electromagnetic interactions, except in general relativity where the electromagnetic field is incorporated into the field equations, are treated independently from one another. Despite the apparent unification of electromagnetic and gravitational fields (i.e. general relativity plus Maxwell's equations) the particles are still described as singularities of the field and therefore the theory inherits all the major difficulties (infinite self-energy and other divergences) of classical electrodynamics. One does, of course, obtain from the field equations the mechanical law of motion of these singularities in the form of the Lorentz's equations of motion. These equations of motion in electrodynamics proper have to be postulated independently from the equations of the electromagnetic field. However, the derivation of the law of motion from general relativity does not even circumvent the difficulty that the field assumes infinite value along the trajectories. Furthermore, just as in classical electrodynamics, the theory does not provide a prescription for the distribution of charge. There is no new idea for the removal of the difficulty associated with the action of the particle's own field on itself, which results in another infinity. Hence the problem of self-energy, even at classical level, remains one of the most important unsolved problems of theoretical physics.

The fundamental premises of a theory sketched above can only be found in the unification of the forces that are well understood with respect to their over all behavior in the asymptotic region where the origin or the structure of the particle is not included. The correct way to incorporate the structure of the elementary particle entails not only restoring a basic symmetry into the description of the electromagnetic field by introducing the axial magnetic current density but one that is associated only with a short range magnetic field. Furthermore, the compatibility of the general covariance of the theory with an extended structure is made possible through an indeterminacy in the distribution of the magnetic charge density in the magnetically neutral core of an elementary particle. All of these, inter alia, will result from the unification of the two most fundamental theories of classical electromagnetic field and the general relativistic theory of gravitation.

Einstein's general relativity accounts for the gravitational field in terms of the curvature of space. Electrodynamics, another example of a long range field, has so far not been formulated on a geometrical basis. One of the motivating ideas in the geometrization of the gravitational forces was Mach's principle according to which the inertial properties of a particle (or more generally of energy) depend on the distribution of matter in the rest of the universe. A further profound observation was the formulation of the principle of equivalence (the equality between inertial and gravitational mass). The principle of equivalence does not, directly, apply to the electromagnetic field except through the gravitational field it produces as a result of its energy density in space. This is the extent of the electromagnetic field's involvement with the principle of equivalence. However it is necessary to account fully for the action of the gravitational field on the electromagnetic field itself. The latter plays an important role in assuring the regularity of the field everywhere. Because of the correspondence principle of the generalized theory of gravitation with respect to general relativity the principle of equivalence and Mach's principle are incorporated into the general frame work of the theory in a natural way.

In the presence of the electromagnetic field the general theory of relativity is based on 10+6 field equations which determine the field variables $g_{\mu\nu}$ and $F_{\mu\nu}$ (the electromagnetic field). We may if we wish neglect the gravitational field and solve what is left (Maxwell's equations) for $F_{\mu\nu}$. In the generalized theory of gravitation the 16 field equations contain the sources of the fields and in turn these sources can only be determined through the knowledge of the fields. The basic physical reality in this theory is the field itself, all other observable quantities are derivable either as functions of these fields or as constants of integration (= constants of the motion) of the field equations.

One of the basic difficulties facing the non-symmetric generalization of general relativity was its physical interpretation in terms of the familiar concepts of physics. This paper contains some progress on the physical implications of the theory. In section II the author's version of the generalized theory of gravitation is summarized in the light of its newly established physical interpretation. In particular the identification of the various quantities and the corresponding physical interpretation of the theory differs entirely from that contained in the earlier papers.[1,2,3] We present a direct assessment of its physical meaning in sections III and IV where we derive the static spherically symmetric forms of the 16 field equations. The first two constants of integration, the electric and magnetic charges, play fundamental roles in the classification of the long and short range forces, in describing electrically neutral matter, and in insuring the regularity of the solutions everywhere. These results and symmetries of the field are discussed in sections IV and V. The electrically and magnetically charged and neutral currents (or polarization currents) and the corresponding fields together with their asymptotic behavior at and near the origin are discussed in section VI. In section VII we give an exact solution for the special case of a spherically symmetric and static field of zero magnetic charge which in the asymptotic limit of large distances reduces to the Nordström solution of general relativity. The same section contains the proof of a theorem on the absence of regular solutions for zero magnetic charge. The contents of the section VIII pertain to the neutral magnetic charge distribution in the case of an elementary particle and concludes with the "magnetic theorem" which relates the magnetic charge to the regularity of the field, to the structure of the elementary particles, to the indeterminacy of the surfaces of zero magnetic charge density in the core of an elementary particle and gives its basic role in the correspondence principle of the theory. In section IX the finite self energy and binding energy of an elementary particle is calculated. This section contains some remarks on the possible cosmological implications of the theory. The paper concludes with section X where a general discussion of the results and also a list of relevant problems for further work have been included. The same section contains a suggestion for the quantization of the theory and for the physical interpretation of the negative energy solutions of the field equations.

II. GENERALIZED THEORY OF GRAVITATION

The theory is based on the nonsymmetric generalization of the symmetric theory (general relativity). The fundamental field variables are the components of the nonsymmetric tensor

$$\hat{g}_{\mu\nu} = \hat{g}_{\{\mu\nu\}} + q^{-1} \hat{g}_{[\mu\nu]} \quad , \qquad (II.1)$$

where the constant q is introduced in order to interpret the anti-symmetric part $\hat{g}_{[\mu\nu]}$ as a generalized electromagnetic field and the symmetric part $\hat{g}_{\{\mu\nu\}}$ as the gravitational field. Thus the constant q has the dimensions of an electric field and will be calculated from the solutions of the field equations for $\hat{g}_{\mu\nu}$. For convenience we shall introduce the tensors $g_{\mu\nu}$ and $\Phi_{\mu\nu}$ by

$$\hat{g}_{\mu\nu} = g_{\mu\nu} + q^{-1} \Phi_{\mu\nu} \quad ,$$

where

$$\hat{g}_{\{\mu\nu\}} = \hat{g}_{\{\nu\mu\}} = g_{\mu\nu} \ , \ \hat{g}_{[\mu\nu]} = - \hat{g}_{[\nu\mu]} = \Phi_{\mu\nu} \ , \qquad (II.2)$$

and where $g_{\mu\nu}$ will assume the role of a metric tensor in space-time.

The tensor $\hat{g}_{\mu\nu}$ is reducible with respect to a transformation of the coordinates since the symmetric and antisymmetric parts of $\hat{g}_{\mu\nu}$ transform separately. This approach of Einstein was critisized by many physicists on the basis that the generalized tensor $\hat{g}_{\mu\nu}$ was reducible and therefore gravitation and electromagnetism were not unified. However, these objections had no physical motivation. If we were to introduce some irreducible quantity to describe both fields as inseparable from one another then we would have to abandon the principle of equivalence and thereby destroy the fundamental premises of general relativity. Thus the reducibility of the generalized quantity $\hat{g}_{\mu\nu}$ is a physical necessity in order to preserve the basic differences between the two long range forces of nature, gravitation and electromagnetism. Furthermore, besides the reducible tensor $\hat{g}_{\mu\nu}$ the generalized theory employs its inverse $\hat{g}^{\mu\nu}$, viz.

$$\hat{g}^{\mu\rho}\hat{g}_{\nu\rho} = \hat{g}^{\rho\mu}\hat{g}_{\rho\nu} = \delta^{\mu}_{\nu} \quad , \tag{II.3}$$

where the contraction of indices are correlated and where the symmetric and anti-symmetric parts of $\hat{g}^{\mu\nu}$ are given by

$$\frac{1}{2}\left(\hat{g}^{\mu\nu} + \hat{g}^{\nu\mu}\right) = \hat{g}^{\{\mu\nu\}} = \frac{g^{\mu\nu}(1+\Omega) - \Phi^{\mu\rho}\Phi^{\nu}{}_{\rho}}{1+\Omega-\Lambda^{2}} \quad , \tag{II.4}$$

$$\frac{1}{2}\left(\hat{g}^{\mu\nu} - \hat{g}^{\nu\mu}\right) = \hat{g}^{[\mu\nu]} = \frac{\Phi^{\mu\nu}-\Lambda f^{\mu\nu}}{1+\Omega-\Lambda^{2}} \quad , \tag{II.5}$$

$$\Omega = \frac{1}{2}\Phi^{\mu\nu}\Phi_{\mu\nu} \;,\; \Lambda = \frac{1}{4}f^{\mu\nu}\Phi_{\mu\nu} \;,\; f^{\mu\nu} = \frac{1}{2\sqrt{(-g)}}\varepsilon^{\mu\nu\rho\sigma}\Phi_{\rho\sigma} \tag{II.6}$$

$$g = \text{Det}\,(g_{\mu\nu})\ ,\ \hat{g} = \text{Det}\,(\hat{g}_{\mu\nu}) = g(1+\Omega-\Lambda^2)\ , \quad \text{(II.7)}$$

and where the associated constant q has been suppressed for economy of notation. The quantities $\sqrt{(-g)}\varepsilon_{\mu\nu\rho\sigma}$ and $\frac{1}{\sqrt{(-g)}}\varepsilon^{\mu\nu\rho\sigma}$ transform as general tensors of fourth rank where $\varepsilon_{\mu\nu\rho\sigma}$ is the usual Levi-Cevita tensor. We may also define a tensor density by

$$\hat{\mathfrak{g}}^{\mu\nu} = \sqrt{(-\hat{g})}\ \hat{g}^{\mu\nu} \quad \text{(II.8)}$$

and a fundamental symmetric tensor

$$b^{\mu\nu} = (B^{-1})^{\mu\nu} = \frac{1}{\sqrt{(-g)}}\,\hat{\mathfrak{g}}^{\{\mu\nu\}} = \frac{g^{\mu\nu}(1+\Omega)-\Phi^{\mu\rho}\ \Phi^{\nu}{}_{\rho}}{\sqrt{(1+\Omega-\Lambda^2)}}$$

$$= \frac{g^{\mu\nu}(1+\frac{1}{2}\Omega)+T^{\mu\nu}}{\sqrt{(1+\Omega-\Lambda^2)}}, \quad \text{(II.9)}$$

$$b_{\mu\nu} = \frac{g_{\mu\nu} + \Phi_{\mu\rho}\ \Phi_{\nu}{}^{\rho}}{\sqrt{(1+\Omega-\Lambda^2)}} = \frac{g_{\mu\nu}(1+\frac{1}{2}\Omega)-T_{\mu\nu}}{\sqrt{(1+\Omega-\Lambda^2)}}\ , \quad \text{(II.10)}$$

where

$$g^{\mu\rho}g_{\nu\rho} = \delta^{\mu}_{\nu}\ ,\ b^{\mu\rho}b_{\nu\rho} = \delta^{\mu}_{\nu}\ ,$$

$$b = \text{Det}(b_{\mu\nu}) = \text{Det}(g_{\mu\nu}) = g\ , \quad \text{(II.10')}$$

$$T_{\mu\nu} = \frac{1}{4}\,g_{\mu\nu}\ \Phi_{\rho\sigma}\Phi^{\rho\sigma} - \Phi_{\mu\rho}\Phi_{\nu}{}^{\rho}\ .$$

The tensor indices will be raised and lowered with the aid of the metric tensor $g_{\mu\nu}$. Thus

$$\Phi^{\mu\nu} = g^{\mu\rho} g^{\nu\sigma} \Phi_{\rho\sigma} .$$

The result (II.10') can be obtained from

$$B = \tilde{K} G^{-1} K$$

where the matrices G, B and K are defined by

$$G = [g_{\mu\nu}] , G^{-1} = [g^{\mu\nu}] , B = [b_{\mu\nu}] , K = [K_{\mu\nu}] ,$$

$$K_{\mu\nu} = \hat{g}_{\mu\nu} (1+\Omega-\Lambda^2)^{-\frac{1}{4}} .$$

Hence

$$\text{Det } B = (\text{Det } K)^2 \text{ Det}(G^{-1}) = g .$$

The fundamental tensor $b_{\mu\nu}$ entails an interesting "scale invariance" property with respect to the time-like vector

$$v^\mu = \frac{dx^\mu}{ds} , v^\mu v_\mu = 1 , \qquad \text{(II.11)}$$

where

$$ds^2 = g_{\mu\nu} dx^\mu dx^\nu .$$

Thus we may define another time-like unit vector V_μ by

$$V_\mu = b_{\mu\nu} v^\nu , V^\mu = b^{\mu\nu} v_\nu ,$$

where

$$v^\mu v_\mu = 1 \quad .$$

By using the property

$$T^\mu_\rho T^\rho_\nu = \delta^\mu_\nu \left(\tfrac{1}{4}\Omega^2+\Lambda^2\right) \quad , \qquad \text{(II.12)}$$

we may write

$$\surd(1+\Omega-\Lambda^2) = \surd[(1+\tfrac{1}{2}\Omega)^2 - p_\mu p^\mu] \quad , \qquad \text{(II.13)}$$

where

$$p_\mu = T_{\mu\nu}v^\nu \quad , \quad p_\mu p^\mu = \tfrac{1}{4}\,\Omega^2+\Lambda^2 \qquad \text{(II.14)}$$

and hence,

$$V_\mu = \frac{(1+\frac{1}{2}\Omega)v_\mu - p_\mu}{\surd[(1+\frac{1}{2}\Omega)^2 - p_\mu p^\mu]} \quad .$$

Hence we see that the square root $\surd(1+\Omega-\Lambda^2)$ has an implicite dependence on the unit vector v_μ and is therefore invariant with respect to replacing the square of $\frac{1}{2}\,T_{\mu\nu}$ (i.e. $\frac{1}{4}\,T_{\mu\nu}T^{\mu\nu}$) by the square of p_μ (i.e. $p_\mu p^\mu$). These relations will be found useful in the discussions of the possible quantization of the theory.

From the above relations we see that the symmetric and anti-symmetric parts of $\hat{g}^{\mu\nu}$ mix the gravitational field tensor $g_{\mu\nu}$ with the generalized electromagnetic field tensor $\Phi_{\mu\nu}$. The $\hat{g}^{\mu\nu}$ is, of course, a reducible tensor and will be used in the action principle of the theory.

The action principle of the theory can be

correlated with that of general relativity. In order to achieve this we shall reformulate the action principle of general relativity by writing

$$S_G = \frac{c^3}{16\pi G} \int L_G \, d^4x \quad , \tag{II.15}$$

where the Lagrangian L_G is given by

$$L_G = \sqrt{(-g)}\, g^{\mu\nu} G_{\mu\nu} + \frac{G}{c^4} \sqrt{(-g)}\; \Phi^{\mu\nu}(\Phi_{\mu\nu} - 2F_{\mu\nu}) \quad , \tag{II.16}$$

and G is the gravitational constant. The second term in the Lagrangian contains the coupling with the strength $\frac{G}{c^4}$ of the electromagnetic field to the gravitational field. The extra variables A_μ in

$$F_{\mu\nu} = \partial_\mu A_\nu - \partial_\nu A_\mu \tag{II.17}$$

are introduced in order to incorporate the special nature (derivability from a potential) of the electromagnetic tensor $\Phi_{\mu\nu}$ into a variational principle. This is a very useful device for the physical interpretation of the various quantities in the generalized theory of gravitation. The action principle of general relativity

$$\delta S_G = 0 \quad , \tag{II.18}$$

applied with respect to the variation of the 20 independent variables $g_{\mu\nu}$, $\Phi_{\mu\nu}$, A_μ leads to the field equations

$$G_{\mu\nu} = \frac{2G}{c^4} T_{\mu\nu} \quad , \qquad \text{(II.19)}$$

$$\frac{\partial}{\partial x^\nu} (\sqrt{(-g)}\ \Phi^{\mu\nu}) = 0 \quad , \qquad \text{(II.20)}$$

$$\Phi_{\mu\nu} = F_{\mu\nu} \quad . \qquad \text{(II.21)}$$

The equation (II.21) implies that the electromagnetic field $\Phi_{\mu\nu}$ satisfies also the remaining Maxwell's equations

$$\Phi_{\mu\nu,\rho} + \Phi_{\nu\rho,\mu} + \Phi_{\rho\mu,\nu} = 0 \quad . \qquad \text{(II.22)}$$

Hence the extra field $F_{\mu\nu}$ is thus eliminated. We must observe that the use of the tensor $\Phi_{\mu\nu}$ for the electromagnetic field in (II.15) must not be confused with its interpretation as "generalized electromagnetic field" in the generalized theory of gravitation where the tensor $\Phi_{\mu\nu}$ refers to the antisymmetric part of $\hat{g}_{\mu\nu}$ and is no longer derivable from a potential.

In order to construct the action principle of the generalized theory in accordance with a correspondence principle we shall rewrite (II.16) in the form

$$L_G = \sqrt{(-g)}(g^{\mu\nu} + q^{-1}\Phi^{\mu\nu})(G_{\mu\nu} - \tfrac{1}{2}\kappa^2 q^{-1} F_{\mu\nu}) + \kappa^2[\tfrac{1}{2}q^{-2}\Omega\sqrt{(-g)}] , \qquad \text{(II.23)}$$

where the constants κ and q are related by

$$\kappa^2 q^{-2} = \frac{4G}{c^4} . \qquad \text{(II.24)}$$

We have thus factorized the coupling constant $\frac{G}{c^4}$ in the Lagrangian (II.16), where the universal constant

$$r_o = \sqrt{2}\, \kappa^{-1} \qquad \text{(II.25)}$$

has the dimensions of a length. Because of the relation (II.24) between κ and q, the Lagrangians (II.23) and (II.16) are equal. We shall give two equivalent Lagrangians for the derivation of the field equations.

METHOD A

The Lagrangian (II.23) can now be generalized by using a one to one correspondence of the form

$$\sqrt{(-g)}(g^{\mu\nu} + q^{-1} \Phi^{\mu\nu}) \rightarrow \hat{g}^{\mu\nu} ,$$

$$G_{\mu\nu} \rightarrow R_{\mu\nu}$$

$$\frac{1}{2} q^{-2} \Omega \sqrt{(-g)} \rightarrow \sqrt{(-\hat{g})} - \sqrt{(-g)}$$

where the expressions on the right hand sides, on expanding and neglecting terms containing powers of q^{-1} higher than 2, reduce to the expressions on the left hand side. Hence the action function for the generalized theory of gravitation, based on the above correspondence, can be expressed as

$$S = \frac{q^2 r_o^2}{8\pi c} \int L \, d^4 x \quad , \tag{II.26}$$

where the Lagrangian L is given by

$$L = \hat{g}^{\mu\nu} \, (R_{\mu\nu} - \tfrac{1}{2} \kappa^2 q^{-1} F_{\mu\nu}) + \kappa^2 [\sqrt{(-\hat{g})} - \sqrt{(-g)}] \quad , \tag{II.27}$$

and where

$$R_{\mu\nu} = - \Gamma^{\rho}_{\mu\nu,\rho} + \Gamma^{\rho}_{\mu\rho,\nu} + \Gamma^{\rho}_{\mu\sigma} \, \Gamma^{\sigma}_{\rho\nu} - \Gamma^{\rho}_{\mu\nu} \, \Gamma^{\sigma}_{\rho\sigma} \tag{II.28}$$

is a "transposition symmetric" curvature tensor. By using the relation

$$\Gamma^{\rho}_{\mu\rho} = \Gamma^{\rho}_{\{\mu\rho\}} = \partial_{\mu} [\mathrm{Ln}\sqrt{(-\hat{g})}] \quad ,$$

the curvature tensor $R_{\mu\nu}$ of the generalized theory can be written as

$$R_{\mu\nu} = -\Gamma^{\rho}_{\mu\nu,\rho} + \partial_{\mu}\partial_{\nu} [\mathrm{Ln}\sqrt{(-\hat{g})}] + \Gamma^{\rho}_{\mu\sigma}\Gamma^{\sigma}_{\rho\nu} - \Gamma^{\rho}_{\mu\nu}\partial_{\rho} [\mathrm{Ln}\sqrt{(-\hat{g})}] . \tag{II.29}$$

The transposition symmetry in this theory corresponds to the charge conjugation invariance of the quantum theory. Thus we may write

$$\hat{g}_{\mu\nu} = (\tilde{\hat{g}})_{\nu\mu} \quad ,$$

where $\tilde{\hat{g}}$ represents the transposed matrix. In a similar way for the nonsymmetric displacement field $\Gamma^{\rho}_{\mu\nu}$ we have

$$\Gamma^{\rho}_{\mu\nu} = (\tilde{\Gamma}^{\rho})_{\nu\mu}$$

where

$$\tilde{\Gamma}^{\rho}{}_{\mu\nu} = \Gamma^{\rho}_{\{\mu\nu\}} - q^{-1} \Gamma^{\rho}_{[\mu\nu]} \quad . \qquad \text{(II.30)}$$

A vector V^{μ} can be displaced parallel to itself by an infinitesimal distance dx^{μ} and the resulting change in its components is given by

$$\delta V^{\rho} = - \Gamma^{\rho}_{\mu\nu} dx^{\mu} V^{\nu}$$

or by its dual displacement

$$\delta V^{\rho} = - (\tilde{\Gamma})^{\rho}_{\mu\nu} dx^{\mu} V^{\nu} \quad .$$

Thus the requirement of transposition invariance of the theory is a necessity to remove this arbitrariness of duality. For the curvature tensor $R_{\mu\nu}$, as seen from its definition (II.28), we have

$$R_{\mu\nu}(\Gamma) = R_{\nu\mu}(\tilde{\Gamma}) \quad , \qquad \text{(II.31)}$$

implying the correlation of the tensor indices μ and ν appearing in the definition (II.28) as first and second indices, respectively. The extra variables A_{μ}, besides maintaining the correspondence with general relativity, play an important role in the transposition invariance of the theory. This role of A_μ will be used more explicitely in the derivation of the field equations from the action principle

$$\delta S = 0 \quad . \tag{II.32}$$

The variation of S with respect to the 16 field variables $\hat{g}_{\mu\nu}$ and four potentials A_μ, as well as with respect to 64 displacement fields $\Gamma^\rho_{\mu\nu}$ leads to the field equations

$$R_{\{\mu\nu\}} = \frac{1}{2}\kappa^2(b_{\mu\nu}-g_{\mu\nu}) \quad , \tag{II.33}$$

$$R_{[\mu\nu]} = \frac{1}{2}\kappa^2(F_{\mu\nu}-\Phi_{\mu\nu}) \quad , \tag{II.34}$$

$$\mathfrak{g}^{[\mu\nu]}{}_{,\nu} = 0 \quad , \tag{II.35}$$

and the transposition invariant algebraic equations

$$\hat{g}_{\mu\nu;\rho} = \hat{g}_{\mu\nu,\rho} - \hat{g}_{\mu\sigma}\,\Gamma^\sigma_{\rho\nu} - \hat{g}_{\sigma\nu}\,\Gamma^\sigma_{\mu\rho} = 0, \tag{II.36}$$

for the $\Gamma^\rho_{\mu\nu}$. The field equations (II.35) result from variation with respect to A_μ. In the absence of (II.35) we would have the result

$$\Gamma^\rho_{[\mu\rho]} = \Gamma_\mu \neq 0$$

and the Lagrangian would not be transposition invariant.

Now, as we did in the field equations of general relativity, we can eliminate the extra field variables $F_{\mu\nu}$ from (II.34) and rewrite the new field equations in the form

$$R_{\{\mu\nu\}} = \frac{1}{2}\kappa^2(b_{\mu\nu}-g_{\mu\nu}) \ , \qquad \text{(II.37)}$$

$$R_{[\mu\nu],\rho} + R_{[\nu\rho],\mu} + R_{[\rho\mu],\nu} + \frac{1}{2}\kappa^2 I_{\mu\nu\rho} = 0 \ , \qquad \text{(II.38)}$$

$$\hat{g}^{[\mu\nu]}{}_{,\nu} = 0 \quad , \qquad \text{(II.39)}$$

where

$$I_{\mu\nu\rho} = \Phi_{\mu\nu,\rho} + \Phi_{\nu\rho,\mu} + \Phi_{\rho\mu,\nu} \quad ,$$

and is an axial 4-vector. Because of the two differential identities obtainable from (II.38) and (II.39) only 16 independent field equations remain to determine 16 field variables $\hat{g}_{\mu\nu}$. The variation with respect to $\hat{g}_{\mu\nu}$ involves the relation

$$\delta[\sqrt{(-g)}] = \frac{1}{2}\, g_{\mu\nu}\, \delta[\sqrt{(-b)}g^{\mu\nu}] = \frac{1}{2}\, b_{\mu\nu}\, \delta\hat{g}^{\mu\nu} \ .$$

The fundamental significance of the extra term involving $F_{\mu\nu}$ in the general relativistic Lagrangian, with a coupling strength $\frac{G}{c^4}$, lies in the fact that without it we could not interrelate or unify, in a physically meaningful way, the fields $g_{\mu\nu}$ and $\Phi_{\mu\nu}$ in the generalized theory.

This is also clear from the simple observation that the Lagrangian (II.27) reduces, in the correspondence limit $r_o = 0$ (or $q = \infty$), to the Lagrangian (II.16) of general relativity. We have thus established a correspondence principle for the generalized theory of gravitation without which the theory could not possibly have a physical basis.

METHOD B

It is interesting to point out that we could, if we wished, obtain the same field equations from the Lagrangian

$$L_o = \hat{g}^{\mu\nu} R_{\mu\nu} + \kappa^2 [\surd(-\hat{g}) - \surd(-g)] \quad , \qquad (II.40)$$

where now the field variables $\hat{g}^{[\mu\nu]}$ are defined according to the equation

$$\hat{g}^{[\mu\nu]} = g^{[\mu\nu\rho]}{}_{,\rho} \qquad (II.41)$$

and where $g^{[\mu\nu\rho]}$ is fully antisymmetric in μ,ν,ρ and is, therefore, an axial 4-vector. Hence the field equations (II.39) are a consequence of the definition (II.41). The potentials $g^{[\mu\nu\rho]}$ can also be defined in the form

$$g^{[\mu\nu\rho]} = \varepsilon^{\mu\nu\rho\sigma} B_\sigma \quad , \qquad (II.42)$$

where the axial 4-vector B_μ generates the field $\hat{g}^{[\mu\nu]}$ according to

$$\Psi_{\mu\nu} = \partial_\mu B_\nu - \partial_\nu B_\mu \quad , \qquad \text{(II.43)}$$

where

$$\Psi_{\mu\nu} = \frac{1}{2}\,\varepsilon_{\mu\nu\rho\sigma}\,\hat{g}^{[\rho\sigma]} \quad . \qquad \text{(II.44)}$$

Hence the field equations (II.39) can be replaced by

$$\Psi_{\mu\nu,\rho} + \Psi_{\nu\rho,\mu} + \Psi_{\rho\mu,\nu} = 0 \quad . \qquad \text{(II.45)}$$

In a similar way, the field equations (II.34) or (II.38) can be stated in the form

$$F_{\mu\nu,\rho} + F_{\nu\rho,\mu} + F_{\rho\mu,\nu} = 0 \quad . \qquad \text{(II.46)}$$

We thus see that the theory contains a vector potential A_μ and an axial vector potential B_μ.

The variation of the Lagrangian (II.40) with respect to the 14 field variables $\hat{g}^{\{\mu\nu\}}$ and B_μ, yield the field equations (II.37) and (II.38).

METHOD C

Now, for the sake of completeness, we shall give a geometrical derivation of the field equations based on the use of the Bianchi-Einstein differential identities

$$\hat{g}^{\mu\nu}(R_{\underset{+-}{\mu\nu};\rho} - R_{\underset{++}{\mu\rho};\nu} - R_{\underset{--}{\rho\nu};\mu}) = 0 \quad , \qquad \text{(II.47)}$$

where the + and - signs signify the correlation

of the indices with respect to the indicated covariant differentiation as first and second indices, respectively. Thus

$$R_{\underset{+-}{\mu\nu};\rho} = R_{\mu\nu,\rho} - R_{\sigma\nu}\Gamma^{\sigma}_{\mu\rho} - R_{\mu\sigma}\Gamma^{\sigma}_{\rho\nu} ,$$

$$R_{\underset{++}{\mu\nu};\rho} = R_{\mu\nu,\rho} - R_{\sigma\nu}\Gamma^{\sigma}_{\mu\rho} - R_{\mu\sigma}\Gamma^{\sigma}_{\nu\rho} ,$$

$$R_{\underset{--}{\mu\nu};\rho} = R_{\mu\nu,\rho} - R_{\sigma\nu}\Gamma^{\sigma}_{\rho\mu} - R_{\mu\sigma}\Gamma^{\sigma}_{\rho\nu} .$$

The identities (II.47) are also satisfied by the fundamental tensor $\hat{g}_{\mu\nu}$,

$$\hat{g}^{\mu\nu}(\hat{g}_{\underset{+-}{\mu\nu};\rho} - \hat{g}_{\underset{++}{\mu\rho};\nu} - \hat{g}_{\underset{--}{\rho\nu};\mu}) = 0 , \qquad \text{(II.48)}$$

where $\hat{g}_{\underset{+-}{\mu\nu};\rho} = 0$, but $\hat{g}_{\underset{++}{\mu\rho};\nu} \neq 0$, $\hat{g}_{\underset{--}{\rho\nu};\mu} \neq 0$.

Furthermore, from multiplying the equation $\hat{g}_{\underset{+-}{\mu\nu};\rho} = 0$ by $g^{\mu\nu}$ we obtain

$$\frac{\hat{g}_{,\rho}}{\hat{g}} - 2\,\Gamma^{\mu}_{\mu\rho} = 0 \quad .$$

Hence

$$[\surd(-\hat{g})]_{;\rho} = [\surd(-\hat{g})]_{,\rho} - \surd(-\hat{g})\;\Gamma^{\mu}_{\rho\mu} = 0 ,$$

and therefore

$$\hat{g}^{\mu\nu}_{\underset{+-}{};\rho} = \hat{g}^{\mu\nu}{}_{,\rho} + \hat{g}^{\sigma\nu}\Gamma^{\mu}_{\sigma\rho} + \hat{g}^{\mu\sigma}\Gamma^{\nu}_{\rho\sigma} - \hat{g}^{\mu\nu}\Gamma^{\sigma}_{\rho\sigma} = 0 . \qquad \text{(II.49)}$$

In the derivation (see Appendix 3) of the

identities (II.47) and (II.48) we have assumed the validity of the equations

$$\hat{g}^{[\mu\nu]}{}_{,\nu} = 0 \quad .$$

Now from the equations (II.49) by contracting with respect to the indices ν and ρ, we obtain

$$\hat{g}^{\{\mu\nu\}}{}_{,\nu} = - \hat{g}^{\rho\sigma}\Gamma^{\mu}_{\rho\sigma} \quad . \qquad (II.50)$$

Hence carrying out the indicated covariant differentiations and using (II.50), we can rewrite the identities (II.47) and (II.48) in the forms

$$[\surd(-g)(b^{\mu\nu}R_{\{\mu\rho\}} - \tfrac{1}{2}\,\delta^{\nu}_{\rho}\; b^{\mu\sigma}R_{\{\mu\sigma\}})]_{\|\nu} =$$

$$\tfrac{1}{2}\,\hat{g}^{[\mu\nu]}[R_{[\mu\nu],\rho} + R_{[\nu\rho],\mu} + R_{[\rho\mu],\nu}] \;, \qquad (II.51)$$

and

$$[\surd(-g)(b^{\mu\nu}g_{\mu\rho} - \tfrac{1}{2}\,\delta^{\nu}_{\rho}b^{\mu\sigma}g_{\mu\sigma})]_{\|\nu} =$$

$$\tfrac{1}{2}\,\hat{g}^{[\mu\nu]}[\Phi_{\mu\nu,\rho}+\Phi_{\nu\rho,\mu}+\Phi_{\rho\mu,\nu}] \;, \qquad (II.52)$$

where the sign ($\|$) refers to covariant differentiation in terms of the Christoffel symbols constructed out of the tensor $b_{\mu\nu}$.

If we set $\Phi_{\mu\nu} = 0$ the identities (II.51) reduce to the general relativistic relation

$$[\surd(-g)(g^{\mu\nu}G_{\mu\rho} - \tfrac{1}{2}\,\delta^{\nu}_{\rho}g^{\mu\sigma}G_{\mu\sigma})]_{|\nu} = 0 \quad .$$

Hence the right hand side of (II.51), in analogy with general relativity, may be representing some kind of force density. Einstein used the identities (II.47) alone in deriving his field equations and was, presumably, not aware of the fact that these identities were also satisfied by the field tensor $\hat{g}_{\mu\nu}$ as well as by the nonsymmetric tensor $b_{\mu\nu}+F_{\mu\nu}$. Einstein assumed that the field equations should stipulate the vanishing of either $R_{\mu\nu}$ or that the vanishing of $R_{\{\mu\nu\}}$ and the way $R_{[\mu\nu]}$ enter (II.47) viz. $R_{[\mu\nu],\rho} + R_{[\nu\rho],\mu} + R_{[\rho\mu],\nu} = 0$. On the other hand Schrödinger assumed the existence of a cosmological constant and proposed the field equations

$$R_{\mu\nu} = \lambda\ \hat{g}_{\mu\nu} \quad .$$

In fact the role of the identities in the derivation of the field equations is more general than the use of $R_{\mu\nu}$ alone. It can easily be seen that the identities (II.47) are invariant under the substiution

$$R_{\mu\nu} \rightarrow R_{\mu\nu}+\lambda\ \hat{g}_{\mu\nu}+\gamma(b_{\mu\nu}+F_{\mu\nu}) \ .$$

This is also clear from (II.48) and also from (II.51) and (II.52). Thus the most general possible form of the field equations is contained in the statement

$$R_{\mu\nu}+\lambda\ \hat{g}_{\mu\nu}+\gamma(b_{\mu\nu}+F_{\mu\nu}) = 0 \quad .$$

Hence we obtain

$$R_{\{\mu\nu\}} = -\lambda\ g_{\mu\nu} - \gamma b_{\mu\nu} \quad ,$$

$$R_{[\mu\nu],\rho} + R_{[\nu\rho],\mu} + R_{[\rho\mu],\nu} = -\lambda\ I_{\mu\nu\rho} \quad .$$

If we assume that for $\Phi_{\mu\nu} = 0$ the field equations must reduce to the field equations of a pure gravitational field without a cosmological constant we obtain

$$\gamma = -\lambda(=-\tfrac{1}{2}\kappa^2)$$

or

$$R_{\{\mu\nu\}} = \tfrac{1}{2}\kappa^2(b_{\mu\nu}-g_{\mu\nu}) \quad ,$$

$$R_{[\mu\nu],\rho}+R_{[\nu\rho],\mu}+R_{[\rho\mu],\nu}+\tfrac{1}{2}\kappa^2 I_{\mu\nu\rho} = 0 \quad .$$

ELECTRIC AND MAGNETIC CURRENTS

We may now, by using the correspondence principle(2), recognize the roles of the various quantities and indentify them for the physical interpretation of the generalized theory. First we observe that the role of the extra variables $F_{\mu\nu}$ in general relativity is clear: they are equal, as a result of the action principle $\delta S_G = 0$, to $\Phi_{\mu\nu}$ whose divergence, because of the absence of charge, vanishes. Hence in the generalized theory we can follow the same path and define an electric current vector by taking the divergence of the anti-symmetric tensor $\Phi_{\mu\nu} + r_o^2 R_{[\mu\nu]}$ $(= F_{\mu\nu})$ and thereby eliminate the extra field variables $F_{\mu\nu}$ (once more) by defining the vector density

$$J^\mu = J_e^\mu + J_o^\mu = \frac{1}{4\pi} \frac{\partial}{\partial x^\nu} (\surd(-g) F^{\mu\nu}) \qquad \text{(II.53)}$$

as the generalized conserved electric current, where

$$J_e^\mu = \frac{1}{4\pi} \frac{\partial}{\partial x^\nu} [\surd(-g) \Phi^{\mu\nu}] \quad , \qquad \text{(II.54)}$$

$$J_o^\mu = \frac{r_o^2}{4\pi} \frac{\partial}{\partial x^\nu} [\surd(-g) R^{[\mu\nu]}] \quad . \qquad \text{(II.55)}$$

In the correspondence limit $r_o = 0$ both of these currents vanish so that the electric currents are consequences of a finite universal length r_o. The subscripts e and o in the definitions (II.54) and (II.55) refer to charged and neutral currents, respectively. The statements

$$\int J_e^\mu \, d\sigma_\mu = \pm e \quad , \quad \int J_o^\mu \, d\sigma_\mu = 0 \quad , \qquad \text{(II.56)}$$

will be proved in section VI. The neutral current J_o^μ will be interpreted as a "polarization current".

The definitions (II.54) and (II.55) imply that knowledge of the electric current J^μ depends on knowledge of the field variables $\hat{g}_{\mu\nu}$ which, in turn, are the solutions of the field equations (II.37)-(II.39). We have now, established the important fact that electric currents are determined according to the laws of the field and that the currents cannot be prescribed arbitrarily.

The theory has also an axial neutral current

$$\Phi_{\mu\nu,\rho} + \Phi_{\nu\rho,\mu} + \Phi_{\rho\mu,\nu} = I_{\mu\nu\rho} = 4\pi\, \varepsilon_{\mu\nu\rho\sigma}\, \mathfrak{s}^{\sigma} ,$$

or

$$\mathfrak{s}^{\mu} = \frac{1}{4\pi} \frac{\partial}{\partial x^{\nu}} [\surd(-g) f^{\mu\nu}] , \qquad \text{(II.57)}$$

where

$$\mathfrak{s}^{\mu} = \surd(-g) s^{\mu} , \quad I^{\mu\nu\rho} = - \frac{1}{\surd(-g)} \varepsilon^{\mu\nu\rho\sigma} s_{\sigma}$$

where the axial vector density $\mathfrak{s}^{\mu}$ is a "magnetic current" and has no classical counterpart since, as seen from the field equations (II.38), in the correspondence limit $r_o = 0$ it vanishes. It is shown in section VIII that $\mathfrak{s}^{\mu}$ represents a magnetically neutral current density (i.e. equal amounts of positive and negative magnetic charge distributions) and vanishes at distances beyond the universal length r_o and that

$$\int \mathfrak{s}^{\mu}\, d\sigma_{\mu} = 0 , \qquad \text{(II.58)}$$

where $d\sigma_{\mu}$ are the 3-dimensional surface elements in the 4-dimensional space. Here again it is clear from the definition (II.57) and from the field equations (II.38) that the distribution of the magnetic charge density is prescribed by the laws of the field and cannot be predetermined. In order to see more explicitly the nature of the magnetic current distribution s^μ we may derive the linearized form of the field equations (II.38) in flat space-time in the form

$$(\nabla^2 - \frac{\partial^2}{c^2 \partial t^2} + \kappa^2) s^\mu = 0 \quad . \qquad \text{(II.59)}$$

This equation is, of course, valid only at distances much larger than r_o, in which case, because of the large size of $\kappa(\sim 10^{14}\,\text{cm}^{-1})$, s^μ is negligibly small. However, the equation does, still, contain useful information on the nature of s^μ. From a plane wave solution $\exp(i\, k_\mu\, x^\mu)$ of (II.59) we see that k^μ is a space-like vector and, therefore, the current s^μ has no wave-like properties and that the magnetic current distribution is confined to distances of the order of r_o. Therefore with respect to the light cone the field $\Phi_{\mu\nu}$ can be decomposed according to

$$\Phi_{\mu\nu} = \Phi^{s}_{\mu\nu} + \Phi^{t}_{\mu\nu} \qquad \text{(II.60)}$$

where

$$\Phi^{s}_{\mu\nu} = (\frac{1}{2\pi})^4 \int \Phi_{\mu\nu}(k) e^{ik_\rho x^\rho} \Theta_-(k) d^4k \ , \qquad \text{(II.61)}$$

$$\Phi^{t}_{\mu\nu} = (\frac{1}{2\pi})^4 \int \Phi_{\mu\nu}(k) e^{ik_\rho x^\rho} \Theta_+(k) d^4k \ , \qquad \text{(II.62)}$$

and

$$\Theta_{-}(k) = \begin{cases} 1 \quad , \quad k_\mu k^\mu < 0 \\ 0 \quad , \quad \text{otherwise} \end{cases} , \qquad \text{(II.63)}$$

$$\Theta_{+}(k) = \begin{cases} 1 \quad , \quad k_\mu k^\mu \geqq 0 \\ 0 \quad , \quad \text{otherwise} \end{cases} \qquad \text{(II.64)}$$

Thus the neutral magnetic current density s^μ contains only the space-like parts of the field $\Phi_{\mu\nu}$.

Now, in addition to the neutral magnetic current density s^μ we may also define a charged magnetic current density ζ^μ as the divergence of the tensor density $\sqrt{(-g)}\Psi^{\mu\nu}$. Thus we write

$$\zeta^\mu = \frac{1}{4\pi} \frac{\partial}{\partial x^\nu} [\sqrt{(-g)}\Psi^{\mu\nu}] \quad , \qquad \text{(II.65)}$$

where, as follows from the spherically symmetric case (see III and IV), we have

$$\int \zeta^\mu \, d\sigma_\mu = \pm g \quad . \qquad \text{(II.66)}$$

The corresponding magnetic fields for the currents s^μ and ζ^μ are short range fields. Therefore a monopole charge is associated with a short range field. It will be shown in sections VIII and IX that the range of the field associated with a monopole is of the order of the nucleon Compton wave length. This result will be interpreted as a classical basis for the strong and weak interactions.

We have thus shown that the four antisymmetric tensors

$$\Phi_{\mu\nu} \ , \ f_{\mu\nu} \ , \ \Psi_{\mu\nu} \ , \ R_{[\mu\nu]} \quad ,$$

of the field provide two electric and two magnetic currents, where $f_{\mu\nu}$ and $r_o^2 \ R_{[\mu\nu]}$ refer to neutral fields.

Now, for the sake of further comparison with the classical theory, we shall cast the field equations (II.38) (or II.45) and (II.39) (or II.46) in a more conventional form in terms of the four generalized electromagnetic field vectors $\mathcal{E}$, $\mathcal{D}$, $\mathcal{H}$, $\mathcal{B}$ defined by

$$\mathcal{E} = [\Phi_{14}+r_o^2 \ R_{[14]} \ , \ \Phi_{24}+r_o^2 \ R_{[24]} \ , \ \Phi_{34}+r_o^2 \ R_{[34]}] ,$$

$$\mathcal{D} = [\hat{g}^{[14]} \ , \ \hat{g}^{[24]} \ , \ \hat{g}^{[34]}] = [\Psi_{23}, \ \Psi_{31}, \ \Psi_{12}]$$

$$\mathcal{H} = [\hat{g}^{[23]} \ , \ \hat{g}^{[31]} \ , \ \hat{g}^{[12]}] = [\Psi_{14}, \ \Psi_{24}, \ \Psi_{34}]$$

$$\mathcal{B} = [\Phi_{23}+r_o^2 \ R_{[23]}, \ \Phi_{31}+r_o^2 \ R_{[31]}, \ \Phi_{12}+r_o^2 \ R_{[12]}] \ ,$$

where in identifying various components of $\Psi_{\mu\nu}$ (II.45) and $F_{\mu\nu}$ (II.46) we have employed the usual polar and axial vector symmetries of these quantities with respect to space-time transformations. Hence the field equations (II.38 or II.46) and (II.39 or II.45) can, in a local Lorentz frame of reference, be written as

$$\nabla \cdot \mathcal{D} = 0 \ , \ \nabla \times \mathcal{H} = \frac{1}{c}\frac{\partial \mathcal{D}}{\partial t} \ , \qquad \text{(II.67)}$$

$$\nabla \cdot \mathcal{B} = 0 \ , \ \nabla \times \mathcal{E} = -\frac{1}{c}\frac{\partial \mathcal{B}}{\partial t} \ , \qquad \text{(II.68)}$$

which, in the correspondence limit $r_o = 0$, reduce to Maxwell's equations for empty space where $\mathcal{B} = \mathcal{H}$, $\mathcal{E} = \mathcal{D}$. The field equations (II.38)-(II.39), when expressed in the form (II.67)-(II.68), resemble the electrodynamics of continuous media. The Φ and R terms in the definition of the generalized electric field $\mathcal{E}$ represent charged and neutral (or polarized) fields, respectively.

For the definitions (II.53) and (II.65) the corresponding equations are given by

$$\nabla \cdot \mathcal{E} = 4\pi(j_o^4 + j_e^4) \ ,$$

$$-\frac{\partial \mathcal{E}}{c\,\partial t} + \nabla \times \mathcal{B} = 4\pi(j_o + j_e) \ ,$$

and

$$\nabla \cdot \mathcal{H} = 4\pi\zeta^4 \ ,$$

$$\frac{\partial \mathcal{H}}{c\,\partial t} + \nabla \times \mathcal{D} = 4\pi\zeta$$

which, of course, are just the definitions of the currents.

In the same way we may rewrite the field equations (II.37) in the form

$$G_{\mu\nu} = \kappa_o^2 \, T_{\mu\nu} \ , \qquad \text{(II.69)}$$

where

$$\kappa_o^2 = \frac{2G}{c^4} \ ,$$

$$R_{\{\mu\nu\}} = - G_{\mu\nu} - S^{\rho}_{\mu\nu;\rho} + S^{\rho}_{\mu\rho;\nu} + \Gamma^{\rho}_{[\mu\sigma]} \Gamma^{\sigma}_{[\rho\nu]} \ ,$$

and where the source term $T_{\mu\nu}$ of the gravitational field is given by

$$- T_{\mu\nu} = \frac{1}{2} \frac{\kappa^2}{\kappa_o^2} (b_{\mu\nu} - g_{\mu\nu}) + \kappa_o^{-2} \Lambda_{\mu\nu} \ , \qquad \text{(II.70)}$$

$$\Lambda_{\mu\nu} = S^{\rho}_{\mu\nu;\rho} - S^{\rho}_{\mu\rho;\nu} - \Gamma^{\sigma}_{[\mu\rho]} \Gamma^{\rho}_{[\sigma\nu]} \ ,$$

$$S^{\rho}_{\mu\nu} = g^{\rho\sigma} \, [\Phi_{\mu\alpha} \Gamma^{\alpha}_{[\sigma\nu]} + \Phi_{\alpha\nu} \Gamma^{\alpha}_{[\mu\sigma]}] \ ,$$

$$\Gamma^{\rho}_{\{\mu\nu\}} = \{^{\rho}_{\mu\nu}\} + S^{\rho}_{\mu\nu} \ , \quad \{^{\rho}_{\mu\nu}\} = \frac{1}{2} g^{\rho\sigma} \, [g_{\mu\sigma,\nu} + g_{\nu\sigma,\mu} - g_{\mu\nu,\sigma}] .$$

The $\Lambda_{\mu\nu}$ term in $T_{\mu\nu}$ is small compared to the first term. The latter in the correspondence limit $r_o \to 0$ reduces to the energy momentum tensor of the electromagnetic field. Hence we see that, just as in general relativity, the generalized theory also can yield Lorentz's equations of motion for point particles[4]. Higher order corrections to these equations of motion are proportional to q^{-2} where q is very large ($\sim 10^{39}$ e.s.u.).

Finally, from the action principle (II.32),

via the Bianchi identities[1] of the nonsymmetric theory, we can derive the conservation laws

$$\mathcal{T}^{\nu}_{\mu,\nu} = 0 \quad , \tag{II.71}$$

where

$$-4\pi\ \kappa^2 q^{-2}\ \mathcal{T}^{\nu}_{\mu} = \hat{g}^{\nu\rho}R_{\mu\rho}+\hat{g}^{\rho\nu}R_{\rho\mu}-\delta^{\nu}_{\mu}\hat{g}^{\rho\sigma}R_{\rho\sigma}+\hat{g}^{\rho\sigma}_{\ ,\mu}\mathcal{B}^{\nu}_{\rho\sigma}-\delta^{\nu}_{\mu}\mathcal{B}, \tag{II.72}$$

$$\mathcal{B}^{\rho}_{\mu\nu} = \delta^{\rho}_{\mu}\Gamma^{\sigma}_{\{\nu\sigma\}}-\Gamma^{\rho}_{\mu\nu} \quad , \quad \mathcal{B} = \hat{g}^{\mu\nu}(\Gamma^{\rho}_{\mu\sigma}\Gamma^{\sigma}_{\rho\nu}-\Gamma^{\rho}_{\mu\nu}\Gamma^{\sigma}_{\{\rho\sigma\}}).$$

On substituting the field equations (II.33)-(II.35) in (II.72) we obtain

$$\mathcal{T}^{\nu}_{\mu} = \frac{1}{4\pi}\ [q^2(\sqrt{(-\hat{g})}-\sqrt{(-g)})\delta^{\nu}_{\mu}-\hat{g}^{[\nu\rho]}F_{\mu\rho}]$$
$$+\ \frac{1}{2}\ \delta^{\nu}_{\mu}\ \mathcal{T}-\ \frac{c^4}{16\pi G}\ \hat{g}^{\rho\sigma}_{\ ,\mu}\ B^{\nu}_{\rho\sigma} \quad , \tag{II.73}$$

where $\mathcal{T} = \mathcal{T}^{\mu}_{\mu}$.

In the limit $r_o \rightarrow 0$ (II.72) and (II.73) reduce to the conventional energy tensor of general relativity. The field equations as well as the energy tensor $\mathcal{T}^{\nu}_{\mu}$ remain invariant under the gauge transformations

$$\Gamma^{\rho}_{\mu\nu} \rightarrow \Gamma^{\rho}_{\mu\nu}+\delta^{\rho}_{\mu}\ \lambda_{,\nu} \quad , \quad A_{\mu} \rightarrow A_{\mu}+\lambda_{,\mu} \quad , \quad B_{\mu} \rightarrow B_{\mu}+\lambda_{,\mu} \ . \tag{II.74}$$

III. STATIC SPHERICALLY SYMMETRIC EQUATIONS

The static spherically symmetric field solutions in general relativity, i.e. the Schwarzschild and Nordström solutions, have provided a satisfactory basis for deriving various physical implications of the theory. It is, therefore, quite natural to adopt the same method for the generalized theory of gravitation. In the latter instance the spherically symmetric field represented by the nonsymmetric tensor $\hat{g}_{\mu\nu}$ has only five nonvanishing components. In order to see this fact we shall discuss the spherically symmetric form of the antisymmetric part $\Phi_{\mu\nu}$. The values of $\Phi_{\mu\nu}$ at, for example, the point $(0,0,z=r,t)$ remain unchanged under the rotation of a local Lorentz frame of reference by an angle $\frac{\pi}{2}$ around the z-axis. The rotation is effected by the matrix

$$R_Z(\tfrac{\pi}{2}) = \exp(i\,\tfrac{\pi}{2}\,M_{12}) \quad , \qquad \text{(III.1)}$$

where the generator of the rotation M_{12} is given by

$$M_{12} = \begin{bmatrix} 0 & -i & 0 & 0 \\ i & 0 & 0 & 0 \\ 0 & 0 & 0 & 0 \\ 0 & 0 & 0 & 0 \end{bmatrix} . \qquad \text{(III.2)}$$

The rotation of the coordinates by $R_Z(\frac{\pi}{2})$ produces the transformations

$$x' = -y \ , \ y' = x \ , \ z' = z \ , \ t' = t \ .$$

Under the transformation (III.1), the tensor $\Phi_{\mu\nu}$ transforms according to the rule

$$[\Phi'_{\mu\nu}] = R_z(\tfrac{\pi}{2}) \ [\Phi_{\mu\nu}]\tilde{R}_z(\tfrac{\pi}{2}) \ . \qquad \text{(III.3)}$$

By using the condition of spherical symmetry

$$\Phi'_{\mu\nu} = \Phi_{\mu\nu} \ , \qquad \text{(III.4)}$$

together with the transformation (III.3) we obtain for the points on the z-axis the results

$$\Phi_{23} = \Phi_{31} = \Phi_{41} = \Phi_{42} = 0 \ .$$

There are thus, for the spherically symmetric field, only two nonvanishing components Φ_{12} and Φ_{43}.

We may now extend the above special transformation and the resulting symmetry obtained to more general transformations pertaining to arbitrary points of space and time. Thus let us apply a new rotation to bring the point (0,0,r,t) to the point (x,y,z,t) where $r^2 = x^2+y^2+z^2$. We first rotate the yz-plane (x=0) around the z-axis by an angle ϕ to coincide with the point (x,y,z,t). The equation of the new plane is

$$a_1 x + a_2 y = 0 \ . \qquad \text{(III.5)}$$

The angle of rotation ϕ is given by

$$\phi = \tan^{-1}(\frac{y}{x}) \ , \qquad \text{(III.6)}$$

and the corresponding rotation matrix has the form

$$R_z(\phi) = \exp\ (i\phi M_{12}) \ . \qquad \text{(III.7)}$$

The explicit form of (III.7) can be obtained by using the relation

$$\exp(i\omega\cdot M) = \exp[i(\omega_1 M_{23} + \omega_2 M_{31} + \omega_3 M_{12})] = 1 + i(\omega\cdot M)\ \frac{\sin\omega}{\omega} + (\omega\cdot M)^2\ (\frac{\cos\omega - 1}{\omega^2}) \ , \qquad \text{(III.8)}$$

where

$$(\omega\cdot M)^3 = \omega^2(\omega\cdot M) \ .$$

The matrix $\exp(i\omega\cdot M)$ represents a rotation by an angle $|\omega|$ around the direction $\hat{\omega} = \omega/\omega$.

We may now perform a rotation by an angle θ in the plane (III.5) around its normal direction $(a_1, a_2, 0)$ to bring the point $(0,0,r,t)$ into the point (x,y,z,t), where

$$\theta = \tan^{-1}[\frac{\sqrt{(x^2+y^2)}}{z}],\ \theta^2 = a_1^2 + a_2^2,\ a_2 = -\frac{x}{y}\ a_1 \ ,$$

or

$$a_1 = \frac{y\theta}{\sqrt{(x^2+y^2)}} \quad , \quad a_2 = -\frac{x\theta}{\sqrt{(x^2+y^2)}} \quad . \qquad \text{(III.9)}$$

Hence the rotation around $(a_1, a_2, 0)$ is effected by the matrix

$$R_a(\theta) = \exp[i(a_1 M_{23} + a_2 M_{31})] \quad . \qquad \text{(III.10)}$$

From the above results it follows that the rotation matrix required to bring the point $(0,0,r,t)$ to the point (x,y,z,t) is given by

$$R = R_a(\theta) R_z(\phi) \quad . \qquad \text{(III.11)}$$

Now, by applying R to the matrix $[\Phi_{\mu\nu}]$ with only two surviving components Φ_{12}, Φ_{34} and then transforming into spherical polar coordinates we obtain the final result

$$[\Phi^S_{\mu\nu}] = SR[\Phi^z_{\mu\nu}]\tilde{R}\tilde{S} = \begin{bmatrix} 0 & 0 & 0 & W \\ 0 & 0 & \chi\sin\theta & 0 \\ 0 & -\chi\sin\theta & 0 & 0 \\ -W & 0 & 0 & 0 \end{bmatrix} , \qquad \text{(III.12)}$$

where for an arbitrary function $f(r)$

$$\chi = r^2 \, f(r) \quad , \qquad \text{(III.13)}$$

and where the matrix S is given by

$$S = \left[\frac{\partial x^\mu}{\partial x'^\nu}\right] \quad , \quad x'^\mu \equiv (r,\theta,\phi,t) \quad ,$$

$x = r\sin\theta\cos\phi$, $y = r\sin\theta\sin\phi$, $z = r\cos\theta$, $t=t'$.

By applying the same steps to the symmetric part $g_{\mu\nu}$ we can construct the most general spherically symmetric tensor in the form

$$[g^S_{\mu\nu}] = \begin{bmatrix} -\dfrac{e^{-u}}{v^2}, & 0, & 0, & \dfrac{1}{v}\tanh\Gamma \\ 0, & -e^{\rho}\sin\Phi, & e^{\rho}\cos\Phi\sin\theta, & 0 \\ 0, & -e^{\rho}\cos\Phi\sin\theta, & -e^{\rho}\sin\Phi\sin^2\theta, & 0 \\ -\dfrac{1}{v}\tanh\Gamma, & 0, & 0, & e^{u} \end{bmatrix} \tag{III.14}$$

where the diagonal elements in agreement with the existence of a light cone are restricted by the condition

$$\sin\Phi \geqq 0 \quad . \tag{III.15}$$

The condition (III.15) on the function $\Phi(r)$ will also be obtained from the field equations which do not have real solutions in the region excluded by (III.15). Furthermore the functional forms of $[\hat{g}^S_{\mu\nu}]$, as will be seen later, for the five functions $u(r)$, $v(r)$, $\Phi(r)$, $\Gamma(r)$, $\rho(r)$ are imposed, in a natural way, by the spherically symmetric forms of the affine connections $\Gamma^{\rho}_{\mu\nu}(\hat{g})$.

In the course of the various manipulations on the field equations we shall need a number of algebraic results and the following is a summary of them. The inverse of $[\overset{S}{g}_{\mu\nu}]$ is given by

$$[\hat{g}_S^{\mu\nu}] = \begin{bmatrix} -v^2 e^u \cosh^2\Gamma, & 0, & 0, & v\cosh\Gamma\sinh\Gamma \\ 0, & -e^{-\rho}\sin\Phi, & -\dfrac{e^{-\rho}\cos\Phi}{\sin\theta}, & 0 \\ 0, & \dfrac{e^{-\rho}\cos\Phi}{\sin\theta}, & -\dfrac{e^{-\rho}\sin\Phi}{\sin^2\theta}, & 0 \\ -v\cosh\Gamma\sinh\Gamma, & 0, & 0, & e^{-u}\cosh^2\Gamma \end{bmatrix} .$$

The determinant of $[\hat{g}^S_{\mu\nu}]$ as follows from (III.14) is given by

$$\hat{g} = \text{Det}\ [\hat{g}^S_{\mu\nu}] = -\frac{e^{2\rho}}{v^2\cosh^2\Gamma}\sin^2\theta ,$$

and that of $[\overset{S}{g}_{\mu\nu}]$ by

$$g = \text{Det}\ [\hat{g}^S_{\{\mu\nu\}}] = -\frac{e^{2\rho}}{v^2}\sin^2\Phi\ \sin^2\theta .$$

Hence

$$\sqrt{(-\hat{g})} = \frac{e^\rho}{v\cosh\Gamma}\sin\theta,\ \sqrt{(-g)} = \frac{e^\rho}{v}\sin\Phi\ \sin\theta . \tag{III.16}$$

We may now write the tensor density $\hat{\mathfrak{g}}^{[\mu\nu]} = \sqrt{(-\hat{g})}\ \hat{g}^{[\mu\nu]}$ as

$$[\hat{g}^{[\mu\nu]}] = \begin{bmatrix} 0, & 0, & 0, & -e^{\rho}\sinh\Gamma\sin\theta \\ 0, & 0, & \frac{\cos\Phi}{v}, & 0 \\ 0, & -\frac{\cos\Phi}{v}, & 0, & 0 \\ e^{\rho}\sinh\Gamma\sin\theta, & 0, & 0, & 0 \end{bmatrix} \tag{III.17}$$

The two fundamental invariants[15] Ω and Λ are

$$\Omega = \frac{1}{2}\,\phi^{\mu\nu}\Phi_{\mu\nu} = \cot^2\Phi - \tanh^2\Gamma \;,$$

$$\Lambda = \frac{1}{4}\,f^{\mu\nu}\Phi_{\mu\nu} = -\cot\Phi\,\tanh\Gamma \;, \tag{III.18}$$

and

$$\sqrt{(1+\Omega-\Lambda^2)} = \frac{1}{\sin\Phi\,\cosh\Gamma} \;.$$

We observe that the tensor density $\hat{g}^{[\mu\nu]}$ as given by (III.17) has some interesting symmetry properties with respect to the transformations $v \to -v$ and $\Gamma \to -\Gamma$. Under these two transformations we obtain $\hat{g}^{[\mu\nu]} \to -\hat{g}^{[\mu\nu]}$. The physical meanings of these transformations are of great significance and are discussed in the section IV.

The spherically symmetric components of the fundamental symmetric tensor $b_{\mu\nu}$ are given by

$$[b^S_{\mu\nu}] = \begin{bmatrix} -\frac{e^{-u}}{v^2}\frac{\sin\Phi}{\cosh\Gamma}, & 0, & 0, & 0 \\ 0, & -e^{\rho}\cosh\Gamma, & 0, & 0 \\ 0, & 0, & -e^{\rho}\cosh\Gamma\sin^2\theta, & 0 \\ 0, & 0, & 0, & \frac{e^{u}\sin\Phi}{\cosh\Gamma} \end{bmatrix} \tag{III.19}$$

In this case also the restriction (III.15) on the function $\Phi(r)$ is an essential requirement.

IV. STATIC SPHERICALLY SYMMETRIC FORM OF THE FIELD EQUATIONS

We shall now discuss the first integrals of the field equations

$$R_{[\mu\nu],\rho} + R_{[\nu\rho],\mu} + R_{[\rho\mu],\nu} + \frac{1}{2}\kappa^2 I_{\mu\nu\rho} = 0 ,$$

$$g^{[\mu\nu]}{}_{,\nu} = 0 ,$$

which for the static spherically symmetric field variables, as follows from Appendix 1, reduce to

$$R_{[23],1} + \frac{1}{2}\kappa^2 \Phi_{23,1} = 0 \tag{IV.1}$$

$$\hat{g}^{[41]}{}_{,1} = 0 . \tag{IV.2}$$

The components $r_o^2 R_{[41]} + \Phi_{[41]}$ and $\hat{g}^{[23]}$ satisfy the field equations identically. The equation (IV.2) can be integrated once and we obtain

$$\hat{g}^{[41]} = \text{const. } \sin\theta \ .$$

From (III.17) we may write

$$e^{\rho} \sinh\Gamma = \pm \lambda_o^2 \ , \qquad \text{(IV.3)}$$

where the constant of integration λ_o^2 can be expressed as

$$\lambda_o^2 = e\, q^{-1} \ . \qquad \text{(IV.4)}$$

The universal constant q being positive, the constant e represents a positive electric charge. The ± signs in (IV.3) are due to the invariance of the equations under the transformation $\Gamma \to -\Gamma$. The generalized electric field E is given by

$$E = \Phi_{14} + r_o^2 R_{[14]} \ ,$$

or

$$E = \frac{1}{v} q \tanh\Gamma$$

$$\left[1+e^{-\rho} r_o^2 \left(v(ve^{u+\rho}\rho'\tanh\Gamma)'\coth\Gamma + \tfrac{1}{2}v^2 e^{u+\rho}(\rho'^2+\Phi'^2)\right)\right] \qquad \text{(IV.5)}$$

where $q \tanh\Gamma = \dfrac{\pm e}{\sqrt{(e^{2\rho}+\lambda_o^4)}}$. Hence we see that the theory predicts two signs for the electric charge.

The first integration of the field equation (IV.1) yields the result

$$r_o^2 \, R_{[23]} + q^{-1} \, \Phi_{23} = \pm \, \ell_o^2 \sin\theta \;, \qquad \text{(IV.6)}$$

where

$$q^{-1} \, \phi_{23} = e^{\rho} \cos\Phi \, \sin\theta$$

and

$$\mathcal{B}_{\phi} = \Phi_{23} = q e^{\rho} \cos\Phi \, \sin\theta \quad ,$$

represents neutral magnetic field, and where the negative sign in (IV.6) can be understood by observing that if $\Phi(r)$ is a solution of (IV.6) corresponding to ℓ_o^2 then $\pi - \Phi(r)$ is another solution corresponding to $- \ell_o^2$.

In view of the axial nature of the left hand side of (IV.6), the constant of integration ℓ_o^2 in (IV.6), can be related to a magnetic charge g by writing ℓ_o^2 in the form

$$\ell_o^2 = g \, q^{-1} \quad , \qquad \text{(IV.7)}$$

where, as in the case of the electric charge appearing in (IV.4), the constant g represents a positive magnetic charge. It will be shown in section VIII that the constant g assumes a spectrum of values. The intrinsic charged magnetic field generated by the charge g is given by

$$\mathcal{H}_g = q\, g^{[23]} = \frac{q}{v} \cos\Phi \quad . \tag{IV.8}$$

Because of the appearance of the function $\cos\Phi$ in the definitions of $\mathcal{B}_o$ and $\mathcal{H}_g$ they represent short range fields.

Now, the remaining field equations to be integrated are

$$R_{11} = \frac{1}{2} \kappa^2 \frac{e^{-u}}{v^2} (1 - \frac{\sin\Phi}{\cosh\Gamma}) \quad , \tag{IV.9}$$

$$R_{22} = \frac{1}{2} \kappa^2 e^{\rho} (\sin\Phi - \cosh\Gamma) \quad , \tag{IV.10}$$

$$R_{33} = R_{22} \sin^2\theta \quad ,$$

$$R_{44} = -\frac{1}{2} \kappa^2 e^{u} (1 - \frac{\sin\Phi}{\cosh\Gamma}) \quad , \tag{IV.11}$$

$$R_{[23]} = \frac{1}{2} \kappa^2 (\ell_o^2 - e^{\rho} \cos\Phi) \sin\theta \quad . \tag{IV.12}$$

By taking the linear combinations

$$R_{[23]} \sin\Phi + R_{22} \cos\Phi \sin\theta \quad ,$$

$$R_{[23]} \cos\Phi - R_{22} \sin\Phi \sin\theta \quad ,$$

$$R_{11} - \frac{e^{-2u}}{v^2} R_{44} \quad ,$$

$$R_{11} + \frac{e^{-2u}}{v^2} R_{44} \quad ,$$

we can achieve considerable simplification and write the nonlinear differential equations in the form,

$$v(ve^{u+\rho}\Phi')'+v^2\Phi'\rho' e^{u+\rho}\tanh^2\Gamma = \kappa^2(e^{\rho}\cos\Phi\ \cosh\Gamma\mp\ell_o^2\sin\Phi)-2\cos\Phi\equiv 2X \ , \tag{IV.13}$$

$$v(ve^{u+\rho}\rho')'+v^2\rho'^2 e^{u+\rho}\tanh^2\Gamma = \kappa^2[e^{\rho}(1-\sin\Phi\ \cosh\Gamma)\mp\ell_o^2\cos\Phi]+2\sin\Phi\equiv 2Y \ , \tag{IV.14}$$

$$v(ve^{u+\rho}u')'+v^2e^{u+\rho}\tanh^2\Gamma(8\rho'^2\tanh^2\Gamma+3u'\rho'-3\rho'^2-\Phi'^2) = \kappa^2e^{\rho}(1-\frac{\sin\Phi}{\cosh\Gamma})\equiv 2Z \ , \tag{IV.15}$$

$$\rho'' + \rho'\ \frac{v'}{v} + \frac{1}{2}\ (\rho'^2+\Phi'^2) - \rho'^2\tanh^2\Gamma = 0 \ , \tag{IV.16}$$

where, as follows from (IV.3), we have used the relations

$$\Gamma' = -\ \rho'\ \tanh\Gamma \ , \ \Gamma'' = (\frac{\rho'^2}{\cosh^2\Gamma} - \rho'')\ \tanh\Gamma \ ,$$

and where X, Y, Z abbreviate the right hand sides of the equations (IV.13), (IV.14) and (IV.15). By multiplying through the equations (IV.13)-(IV.15) by $\frac{1}{\cosh\Gamma}$ and using (IV.16) in (IV.15) to eliminate Φ'^2 we may rewrite them in the simple forms as

$$v\left(\frac{ve^{u+\rho}\Phi'}{\cosh\Gamma}\right)' = \kappa^2\left(e^\rho\cos\Phi \mp \frac{\ell_o^2\sin\Phi}{\cosh\Gamma}\right) - 2\,\frac{\cos\Phi}{\cosh\Gamma}\ , \quad \text{(IV.17)}$$

$$v\left(\frac{ve^{u+\rho}\rho'}{\cosh\Gamma}\right)' = \kappa^2\left[e^\rho\left(\frac{1}{\cosh\Gamma} - \sin\Phi\right) \mp \frac{\ell_o^2\cos\Phi}{\cosh\Gamma}\right] + 2\frac{\sin\Phi}{\cosh\Gamma}, \quad \text{(IV.18)}$$

$$v\left[\frac{ve^{u+\rho}}{\cosh\Gamma}(u'+2\rho'\tanh^2\Gamma)\right]' = \frac{\kappa^2e^\rho}{\cosh\Gamma}\left(1 - \frac{\sin\Phi}{\cosh\Gamma}\right)\ . \quad \text{(IV.19)}$$

$$\rho'' + \rho'\,\frac{v'}{v} + \frac{1}{2}\,(\rho'^2+\Phi'^2) - \rho'^2\tanh^2\Gamma = 0\ . \quad \text{(IV.20)}$$

In the equations (IV.17) and (IV.18) we retain $-\ell_o^2$ on the right hand sides for the solution $\Phi(r)$ and ℓ_o^2 for the solution $\pi-\Phi(r)$. Thus equations (IV.17)-(IV.18) remain invariant under the transformations

$$\Phi(r) \to \pi - \Phi(r)\ ,\ \ell_o^2 \to -\ell_o^2\ . \quad \text{(IV.21)}$$

In general there exist two classes of solutions

$$\mp 2n\pi + \Phi(r) = f_n^+\ , \quad \text{(IV.22)}$$

with positive magnetic charge in the future light cone [corresponding to retaining $-\ell_o^2$ in (IV.17) and (IV.18)] and

$$\mp (n + \tfrac{1}{2})2\pi - \Phi(r) = f_n^-\ , \quad \text{(IV.23)}$$

with the negative magnetic charge in the past light cone [corresponding to retaining $+\ell_o^2$ in (IV.17) and (IV.18)], where

$$n = 0, 1, 2, \ldots \quad .$$

Hence we see that the theory predicts, for the neutral field $\mathcal{B}_o$, the two signs for the magnetic charge simultaneously with the corresponding two sets of infinite number of solutions. These solutions, because of the nonlinearity of the equations, represent nontrivial solutions of the field equations. Both signs of the magnetic charge, in contrast to the electric charge, must for the field $\mathcal{B}_o$ occur at the same time. For the field $\mathcal{B}_o$ the two types of magnetic charges ($\pm g$) are not separable. Furthermore, the solutions (IV.22) and (IV.23) because of the relations $\sin[\pm 2n\pi+\Phi] = \sin\Phi$, $\sin\,[\pm(2n+1)\pi-\Phi] = \sin\Phi$, are consistent with requirement (III.15) on the metrical coefficients of the field. There are no solutions for $\sin(\Phi\pm\frac{\pi}{2}) = \pm\cos\Phi$, except when $\Phi = \frac{\pi}{2}$. The latter possibility is discussed in section VII.

The other two fundamental symmetries of the field equations refer to invariance under the transformations of electric and magnetic charge conjugation

$$v(r) \rightarrow -\,v(r) \quad , \qquad \text{(IV.24)}$$

and electric charge reflection

$$\Gamma(r) \rightarrow -\,\Gamma(r) \quad . \qquad \text{(IV.25)}$$

The invariance under (IV.24) describes, as follows from the definitions of the electric and magnetic fields by (IV.5) and (IV.8), both electric and magnetic charge conjugation. In fact (see section IX) under (IV.24) the energy of the field also changes sign. Therefore the symmetry (IV.24) predicts the existence of particle (positive energy) and antiparticle (negative energy) pairs. The necessary requirement of positivity of the energy for particles and antiparticles will, presumably, be achieved by a possible application of quantum field theory or by some other procedure to be discovered.

The symmetry (IV.25) implies merely the existence of two signs for the electric charge which fact is contained explicitly in the definitions of the electric and magnetic fields and implicitly in the equations (IV.17) - (IV.20). The symmetry (IV.25) does not effect the sign of the magnetic charge. If we apply both transformations (IV.24) and (IV.25) then the electric charge does not change its sign but the sign of the energy and the sign of the magnetic charges change. This fact implies the existence of antiparticles with positive or negative electric and magnetic charges. However if $\Gamma = 0$ then the symmetry (IV.24) predicts the existence of electrically neutral particle, antiparticle pairs. There are no solutions with $v = 0$. The magnetic field H_g under (IV.24) goes to $-H_g$ and the latter under (IV.23) (time reflection) is restored back to the original field.

All elementary particles carry a net magnetic

charge of positive or negative sign associated with a short range field. Particles and anti-particles carry equal but opposite sign magnetic charges superimposed over a magnetically neutral core.

V. SPECIAL SOLUTIONS OF THE FIELD EQUATIONS

Let us begin by solving the field equations (IV.13) and (IV.14) for $\cos\Phi$ and $\sin\Phi$ in terms of X and Y. Thus, putting

$$\sqrt{(e^{2\rho}+\lambda_o^4)} - r_o^2 = R^2 \quad , \tag{V.1}$$

and using $\sin^2\Phi + \cos^2\Phi = 1$ we obtain

$$\cos\Phi = \frac{\pm\ell_o^2\cos\alpha+R^2\sin\alpha}{\sqrt{(R^4+\ell_o^4)}} \quad , \quad \sin\Phi = \frac{\mp\ell_o^2\sin\alpha+R^2\cos\alpha}{\sqrt{(R^4+\ell_o^4)}} \quad , \tag{V.2}$$

where

$$X\, r_o^2 = \sqrt{(R^4+\ell_o^4)}\, \sin\alpha \quad , \tag{V.3}$$

$$e^{\rho}-r_o^2 Y = \sqrt{(R^4+\ell_o^4)}\, \cos\alpha \quad , \tag{V.4}$$

$$\tan\,(\alpha + \Phi) = \pm\,\frac{R^2}{\ell_o^2} \quad , \tag{V.5}$$

and $\alpha(r)$ is a function of r.

The magnetic field H_g can now be written as

$$H_o = \frac{q}{v}\cos\Phi = \frac{\pm g}{v}\,\frac{\cos\alpha \pm \frac{R^2}{\ell_o^2}\sin\alpha}{\surd(R^4+\ell_o^4)} \quad . \tag{V.6}$$

For $\Phi(r)$ = constant we can solve the equations (IV.17) and (IV.18) to obtain

$$\tan\Phi = \frac{\ell_o^4+\lambda_o^4-r_o^4}{2r_o^2} \quad , \tag{V.7}$$

$$\surd(e^{2\rho}+\lambda_o^4)-r_o^2 = \frac{\ell_o^4+\lambda_o^4-r_o^4}{2r_o^2} \quad , \tag{V.8}$$

where we used the definition

$$\coth\Gamma = e^{-\rho}\surd(e^{2\rho}+\lambda_o^4) \quad .$$

For $\Phi(r) = \pm n\pi$ we obtain

$$e^{\rho} = \ell_o^2 \ , \ r_o^2 = \surd(\ell_o^4+\lambda_o^4) = q^{-1}\surd(e^2+g^2) \ . \tag{V.9}$$

Thus, using the relation

$$q^2 r_o^2 = \frac{c^4}{2G}$$

we get the results

$$r_o^2 = \frac{2G}{c^4}(e^2+g^2) \ , \ q = \frac{\surd(e^2+g^2)}{r_o^2} \quad . \tag{V.10}$$

Now by using the equation (IV.17) we can

discuss the stationary behavior of the function $\Phi(r)$. Thus if we set

$$\Phi'(r) = 0 ,$$

then from (IV.17) we obtain

$$v^2 \frac{e^{u+\rho}}{\cosh\Gamma} \Phi'' = \kappa^2 (e^{\rho} \cos\Phi \pm \frac{\ell_o^2 \sin\Phi}{\cosh\Gamma}) - \frac{2\cos\Phi}{\cosh\Gamma} .$$

The nature of the stationary points for the function $\Phi(r)$ will depend on the sign of $\Phi''(r)$. From (V.8) we see that the solutions

$$\Phi(r) = \pm n\pi , \quad n=0,1,2,\ldots$$

correspond to the point $r=0$ where $\Phi'(0) = 0 = \Phi''(0)$. Hence the origin is an inflexion point. The theory, so far, does not relate e and g. In order to find a relation between e and g we shall need an additional requirement, namely introduction of the constant $\hbar$ by a quantization of this theory. An important dimensionless number is the ratio

$$\frac{\ell_o^2}{\lambda_o^2} = \frac{g}{e} = f^2 \qquad \text{(V.11)}$$

where the constant f, as will be seen, as a function of mass is a measure of the strength of the coupling between the various regions of the field at distances of the order of ℓ_o from the origin.

The three lengths

$$\lambda_o^2 = \frac{2G}{c^4} e\sqrt{(e^2+g^2)} \quad , \qquad (V.12)$$

$$\ell_o^2 = \frac{2G}{c^4} g\sqrt{(e^2+g^2)} \quad , \qquad (V.13)$$

$$r_o^2 = \frac{2G}{c^4} (e^2+g^2) \quad , \qquad (V.14)$$

are related according to

$$\lambda_o \leqq \ell_o \leqq r_o \quad . \qquad (V.15)$$

The lengths λ_o and ℓ_o may serve to differentiate between leptonic and hadronic processes, respectively. In (V.15) the equality $\lambda_o = \ell_o$ holds only for $g = e$. For $\ell_o = r_o$ we must set, for an arbitrary g, $e = 0$.

On using (V.9) and (V.10) in the definition (V.1) we obtain $R^2 = 0$. Therefore for the solutions $\Phi(r) = \pm n\pi$ we have $\tan(\alpha+\Phi) = 0$ or $\alpha(r) = \pm n\pi$. Furthermore for $R^2 < \ell_o^2$ the equation (V.2) yields

$$\sin\phi \sim \frac{R^2}{\ell_o^2} \quad , \qquad (V.16)$$

which reconfirms the statement (III.15). On the other hand for $\Phi(r) = \frac{\pi}{2}$ (and therefore $\Phi' = 0$) the equation (IV.13) yields the result $\ell_o^2 = q^{-1}g = 0$ which for $q = \infty$ produces the field equations of general relativity and Maxwell's equations. However for $q = \infty$ we have $\lambda_o^2 = 0$. In this case the equation (V.1) reduces to

$$e^{\rho} = R^2 = r^2 \quad . \tag{V.17}$$

Hence for $R^2 >> \ell_o^2$ equation (V.2) gives the result

$$\cos\Phi \sim \frac{\pm\ell_o^2}{R^2} \; , \; \sin\Phi \sim 1 \quad . \tag{V.18}$$

It will be proved in section VIII that in general we have the relation

$$e^{\rho} = \sqrt{(r^4+\ell_o^4)} \quad . \tag{V.19}$$

Thus the function R as defined by (V.1) plays[16] the role of an "effective radius" and in terms of r is given by

$$R^2 = \sqrt{(r^4+\ell_o^4+\lambda_o^4)} - r_o^2 \quad . \tag{V.20}$$

We may now, formally, solve the equation (IV.20) in the form

$$\frac{1}{v} = \pm \frac{r^3\sqrt{(r^4+\ell_o^4+\lambda_o^4)} \; \exp(F)}{(r^4+\ell_o^4)^{5/4}} \tag{V.21}$$

where

$$F = \frac{1}{2} \int \frac{\Phi'^2}{\rho'} \, dr \quad . \tag{V.22}$$

Let us now consider the asymptotic region

$$r << \ell_o \;\; (f>1) \quad .$$

In this case, using (V.9), we may write

$$R \sim \frac{r^2}{r_o\sqrt{2}} \quad . \tag{V.23}$$

From (V.16), for $r \ll \ell_o$, we get the approximations

$$\Phi' \sim \frac{2r^3}{r_o^2 \ell_o^2} \tag{V.24}$$

$$e^{\rho} \sim \ell_o^2 + \frac{r^4}{2\ell_o^2} \quad , \tag{V.25}$$

$$e^{-\rho} \sim \frac{1}{\ell_o^2}\left(1 - \frac{r^4}{2\ell_o^4}\right) \quad ,$$

$$\rho' \sim \frac{2r^3}{\ell_o^4} \tag{V.26}$$

$$\sqrt{(e^{2\rho}+\lambda_o^4)} \sim r_o^2 + \frac{r^4}{2r_o^2} \quad ,$$

$$\cosh\Gamma \sim \frac{r_o^2}{\ell_o^2} + \frac{1}{2}\,\frac{r^4}{r_o^2\ell_o^2} - \frac{r^4 r_o^2}{2\ell_o^6} \quad , \tag{V.27}$$

$$\sinh\Gamma \sim \pm\, f^{-2}\left(1 - \frac{r^4}{2\ell_o^4}\right) \quad , \tag{V.28}$$

$$\tanh\Gamma \sim \pm\, \frac{\lambda_o^2}{r_o^2}\left(1 - \frac{r^4}{2r_o^4} + \frac{r^4}{2\ell_o^4}\right) \ , \tag{V.29}$$

$$F \sim \frac{r^4}{4r_o^4} \tag{V.30}$$

$$\frac{1}{v} \sim \pm \frac{r^3 r_o^2}{\ell_o^5} \exp\left(\frac{r^4}{4r_o^4}\right) \quad . \tag{V.31}$$

In the asymptotic region where $r \ll \ell_o$ the gravitational potential, which results, from putting $\rho' = \Phi' = 0$ in equation (IV.19), is given by

$$e^u \sim 1 + \kappa^2 \int \frac{dr}{v} \left(\int \frac{dr}{v}\right) \tag{V.32}$$

$$\sim 1 + \left(\frac{r_o}{\ell_o}\right)^{10} \exp(2F) \ , \tag{V.33}$$

where, because of the symmetry of equation (IV.19) under the substitution $v \to -v$ one of the constants of integration associated with the term $\int \frac{dr}{v}$ is set equal to zero. The appearance of the exponential factor exp(2F), since F is positive, indicates the long range character of the gravitational force even at distances where $r \ll \ell_o$.

On dividing both sides of (IV.19) by v and substituting (V.32) for exp(u) we easily see that it is satisfied at r=0 as were equations (IV.17) and (IV.18). At r=0 we obtain the exact result

$$\lim_{r=0} e^u = 1 + \frac{1}{c^2} V_G \ , \tag{V.34}$$

where

$$V_G = \left(\frac{r_o}{\ell_o}\right)^{10} c^2 \quad , \qquad (V.35)$$

represents the value of the gravitational potential at the origin. For the vanishing magnetic charge g it assumes an infinite value. Thus the regularity of the gravitational field everywhere is due to the fact that $g \neq 0$. The mass dependence of (V.35) follows from the possible nature of the magnetic charge g. According to this theory a particle is created by a "gravitational condensation" of the electromagnetic energy density by its own gravitational field and therefore it is quite natural to expect the mass of a particle to depend, among other things, on g. This is also expected from the short range character of the magnetic field due to magnetic charge g.

The space-time line element ds^2 defined in terms of the metrical coefficients $g_{\mu\nu}(= g_{\{\mu\nu\}})$ has the form

$$ds^2 = g_{\mu\nu}dx^\mu dx^\nu = c^2 e^u dt^2 - e^\rho \sin\Phi(d\theta^2 + \sin^2\theta d\phi^2) - \frac{e^{-u}}{v^2} dr^2 \ . \qquad (V.36)$$

At the origin we obtain the time-like line element

$$ds_o^2 = c^2 \left[1 + \frac{1}{c^2} V_G\right] dt^2 \quad . \qquad (V.37)$$

VI. ELECTRIC CHARGE DISTRIBUTION

For the present case of a spherically symmetric static field the only surviving component of the electric current defined by (II.53) is the charge density J^4, viz.,

$$J^4 = J^4_e + J^4_o \quad , \qquad \text{(VI.1)}$$

where

$$J^4_e = \frac{q}{4\pi} (e^{\rho}\tanh\Gamma\sin\Phi)' \sin\theta = \frac{\pm e}{4\pi} \left(\frac{\sin\Phi}{\cosh\Gamma}\right)' \sin\theta \ , \qquad \text{(VI.2)}$$

$$J^4_o = \frac{qr_o^2}{4\pi} [\tfrac{1}{2} v^2 e^{u+\rho}(\rho'^2+\Phi'^2) \tanh\Gamma$$

$$+ v(v\ e^{u+\rho}\rho'\tanh\Gamma)']'\sin\theta \quad , \qquad \text{(VI.3)}$$

$$= \frac{\pm er_o^2}{4\pi} [\frac{\sin\Phi}{\cosh\Gamma} \rho' v^2 e^u \left(Ln \left(\frac{e^u}{\cosh^2\Gamma}\right)\right)']' \sin\theta \quad ,$$

and where we have employed the relations (A_1.7), (IV.3), (IV.16) and

$$R_{[14]} = v\ \Gamma'\ e^u \left[Ln \left(\frac{e^u}{\cosh^2\Gamma}\right)\right]' \quad , \qquad \text{(VI.4)}$$

$$q\sqrt{(-g)}R^{[14]} = \pm e\rho' v^2 \frac{e^u \sin\Phi}{\cosh\Gamma} \left[Ln \left(\frac{e^u}{\cosh^2\Gamma}\right)\right]' \sin\theta \ . \qquad \text{(VI.5)}$$

The neutral charge density[9] J^4_o depends on the

gravitational potential e^u and therefore on the mass itself. The definitions J_e^4 and J_o^4 do, of course, satisfy the conservation laws (II.56), since

$$Q_e = \int J_e^4 \, dr \, d\theta \, d\phi = \pm \, e \, [\frac{\sin\Phi}{\cosh\Gamma}]_o^\infty = \pm \, e \, , \qquad \text{(VI.6)}$$

$$Q_o = \int J_o^4 \, dr \, d\theta \, d\phi =$$

$$\pm \, e \, r_o^2 \, [\frac{\sin\Phi}{\cosh\Gamma} \, \rho' v^2 e^u (Ln(\frac{e^u}{\cosh^2\Gamma}))']_o^\infty = 0 \, , \qquad \text{(VI.7)}$$

where we have used the equations (V.26) - (V.31).

A neutral particle can be defined by setting $J_e^4 = 0$ or, as follows from (VI.2), by taking

$$\sin\Phi = A \cosh\Gamma \qquad \text{(VI.8)}$$

where A is a constant. We observe that an electrically neutral particle, as defined by (VI.8) does still carry a polarization charge density and the latter is given by (VI.3) where $\sin\Phi/\cosh\Gamma$ is replaced by the constant A.

The presence of the neutral charge density implies a structure for a spherically symmetric elementary particle. For $r_o = 0$ the predicted structure reduces to the point description of the conventional theory. The charge densities J_e^4 and J_o^4 are derived from the generalized electric field

$$E = \Phi_{14} + r_o^2 \, R_{[14]} = E_e + E_o \quad , \qquad \text{(VI.9)}$$

where

$$E_e = \frac{1}{v} \frac{\pm e}{R^2+r_o^2} \quad , \qquad (VI.10)$$

$$E_o = \frac{(\pm e)r_o^2}{R^2+r_o^2} \rho' v e^u \; [Ln(\frac{e^u}{\cosh^2\Gamma})]' \quad , \qquad (VI.11)$$

and they vanish at the origin r=0. The calculation of the electric current J^{μ} depends on knowledge of the field variables $\Phi_{\mu\nu}$ which, in turn, are the solutions of the field equations. Thus the electric current cannot be prescribed arbitrarily but is determined from the field itself. The dependence of E_e on $\frac{1}{v}$ and of E_o on v is related, as will be seen, to the range of these fields. The quantity E_e represents a long range field regular everywhere which is due to a positive or negative charge density. The field E_o is regular everywhere and represents a neutral short range field caused by a neutral charge distribution. For $\Phi = \frac{\pi}{2}$ (which corresponds to g=0) equation (VI.10), as follows from (V.21), reduces to the Coulomb[10] field $\pm e/r^2$.

In order to see the nature of the above physical quantities more explicitly we must obtain their asymptotic forms for the two regions $r < \ell_o$ and $r >> r_o$. First let us consider the asymptotic behavior near the origin where, as follows from (V.23) - (V.32), one obtains the results

$$J_e^4 \sim \frac{\pm e}{2\pi} \frac{r^3}{r_o^4} \sin\theta \quad , \qquad (VI.12)$$

$$J_o^4 \sim \frac{\pm e}{\pi} \frac{2r^3}{\ell_o^4} \sin\theta \quad , \qquad \text{(VI.13)}$$

$$E_e \sim \pm\, e\, \frac{(\pm 1)\, r^3 \exp(F)}{\ell_o^5} \quad , \qquad \text{(VI.14)}$$

$$E_o \sim \pm\, e\, (\pm 1)\, \frac{4r^3}{\ell_o^5} \left(\frac{r_o}{\ell_o}\right)^4 \exp(F) \;, \qquad \text{(VI.15)}$$

all of which vanish at the origin provided $g \neq 0$ or $q^{-1} \neq 0$. The numerical factor (± 1) in (VI.14) and (VI.15) results from particle and antiparticle conjugation.

VII. SPECIAL EXACT SOLUTIONS

In order to assess fully the significance of the magnetic charge in this theory we shall study the solutions of equations (IV.13) - (IV.15) for the special case

$$\Phi(r) = \frac{\pi}{2} \quad . \qquad \text{(VII.1)}$$

This result follows also from setting $g = 0$ in the relation (V.6). On substituting $\Phi = \frac{\pi}{2}$, $\Phi' = 0$ in the equation (IV.12) we obtain

$$\ell_o^2 = 0 \quad . \qquad \text{(VII.2)}$$

Hence for the class of solutions (VII.1) the magnetic charge g must vanish so that these solutions are valid only beyond the spectrum of g-values and beyond the distances where $g = 0$. Magnetically neutral surface implied by (VII.1) is, in view of the unknown value of r, indeterminate. This is a consequence of general covariance according to which it is not possible to have a sharply defined rigid object. From the definitions (V.19), (V.22) and (IV.3) we obtain

$$e^{\rho} = r^2 \ , \ v = \pm \frac{r^2}{\sqrt{(r^4+\lambda_o^4)}}, \ \tanh\Gamma = \pm \frac{\lambda_o^2}{\sqrt{(r^4+\lambda_o^4)}} \ . \tag{VII.3}$$

The equations (IV.17) and (IV.20) are satisfied identically and equations (IV.18), (IV.19) reduce to

$$v\left(\frac{ve^{u+\rho}\rho'}{\cosh\Gamma}\right)' = \kappa^2[e^{\rho}(\frac{1}{\cosh\Gamma} - 1)] + \frac{2}{\cosh\Gamma} \ , \tag{VII.4}$$

$$v\left[\frac{ve^{u+\rho}}{\cosh\Gamma}(u'+2\rho'\tanh^2\Gamma)\right]' = \frac{\kappa^2 e^{\rho}}{\cosh\Gamma}(1 - \frac{1}{\cosh\Gamma}) \ . \tag{VII.5}$$

By using (VII.3), equation (VII.4) can be written as

$$\frac{d}{dr}\left(\frac{r^5 e^u}{r^4+\lambda_o^4}\right) = 1 + \frac{1}{2}\kappa^2 \ [r^2-\sqrt{(r^4+\lambda_o^4)}] \ , \tag{VII.6}$$

and can be integrated, at once, in the form

$$e^u = \frac{r^4+\lambda_o^4}{r^4}\left[1-\frac{2Gm}{c^2 r}+\frac{1}{3}\frac{r^2}{\lambda_o^2}-\frac{1}{3}\frac{\sqrt{(r^4+\lambda_o^4)}}{\lambda_o^2}-\frac{2}{3}\frac{\lambda_o^2}{r}K(r)\right], \tag{VII.7}$$

where

$$K(r) = \int\frac{dr}{\sqrt{(r^4+\lambda_o^4)}} = \frac{1}{2\lambda_o}\int\frac{d\gamma}{\sqrt{(1-\frac{1}{2}\sin^2\gamma)}},$$

is an elliptic integral of the first kind and where we have employed the relations

$$\gamma = \cos^{-1}\left(\frac{\lambda_o^2-r^2}{\lambda_o^2+r^2}\right), \quad \kappa^2\lambda_o^2 = 2$$

$$\int\frac{r^4\,dr}{\sqrt{(r^4+\lambda_o^4)}} = \frac{1}{3}\left[r\sqrt{(r^4+\lambda_o^4)}-\lambda_o^4 K(r)\right],$$

$$\int\sqrt{(r^4+\lambda_o^4)}\,dr = \frac{1}{3}\left[2\lambda_o^4 K(r)+r\sqrt{(r^4+\lambda_o^4)}\right].$$

The solution (VII.7), as can be seen by direct substitution, satisfies equation (VII.5). Hence equations (VII.4) and (VII.5) are compatible. The solution (VII.7) is singular at $r = 0$. For the charged part of the electric field we have

$$E_e = \frac{q}{v}\tanh\Gamma = \frac{\pm e(\pm 1)}{r^2}, \tag{VII.8}$$

which is just the usual Coulomb field, where (± 1) as before correspond to particle and antiparticle conjugation. The neutral field can be calculated as

$$E_o = q\lambda_o^2 R_{[14]}$$

$$= \pm e(\pm 1) \frac{4}{3r^2} [1 - \frac{\sqrt{(r^4+\lambda_o^4)}}{r^2} + \frac{\lambda_o^2}{r^3} (\frac{3Gm}{c^2} + \lambda_o^2 K(r))], \tag{VII.9}$$

which falls off as $\frac{1}{r^5}$. The electric charge density

$$J_e^4 = \frac{\pm e}{2\pi} \frac{\lambda_o^4 r}{(r^4+\lambda_o^4)^{3/2}} \sin\theta , \tag{VII.10}$$

for $r \gg \lambda_o$ falls off as $\frac{\lambda_o^4}{r^5}$. The total charge is, of course, conserved since

$$\int J_e^4 \, dr \, d\theta \, d\phi = [\frac{\pm er^2}{\sqrt{(r^4+\lambda_o^4)}}]_o^\infty = \pm e . \tag{VII.11}$$

However, the neutral charge density J_o^4, in view of its singularity at $r = 0$, is not conserved[11]. This illustrates the fact that the neutral charge density is held together by the neutral magnetic charge distribution in the particle itself. The solutions (VII.3) and (VII.7) belong to the asymptotic region beyond the critical and indeterminate point r_c where $\Phi(r_c) = \frac{\pi}{2}$ and $g = 0$.

Now for the asymptotic limit $r \gg \lambda_o$ the solution (VII.7) reduces to

$$e^u \to 1 - \frac{2Gm}{c^2 r} + \frac{Ge^2}{c^4 r^2} , \tag{VII.12}$$

which is Nordström's extension of the Schwarzschild solution of general relativity in the presence of an electric field. The result (VII.12) can not be obtained for the nonvanishing magnetic charge g.

Theorem: There exist no regular solutions where g = 0 and conversely for g = 0 the solutions are not regular everywhere.

In order to prove the theorem let us consider the asymptotic solution for the equations (IV.17) - (IV.19) in the region where $r >> \lambda_o$. From $R \sim r$, $e^{\rho} \sim r^2$ and from equations (V.2) we obtain

$$\cos\Phi \sim \frac{\pm \ell_o^2}{r^2} - \frac{1}{2}\frac{\pm \ell_o^2 r_o^2}{r^4} , \qquad \text{(VII.13)}$$

$$\sin\Phi \sim 1 - \frac{\ell_o^4}{2r^4} , \quad \Phi' \sim \frac{2\ell_o^2}{r^3} .$$

Hence

$$F = \frac{1}{2}\int \frac{\Phi'^2}{\rho'} dr \sim - \frac{\ell_o^4}{4r^4} ,$$

$$\frac{1}{v} \sim \pm \exp(-\frac{\ell_o^4}{4r^4}) \sim \pm (1 - \frac{\ell_o^4}{4r^4}) , \qquad \text{(VII.14)}$$

$$\cosh\Gamma \sim 1 + \frac{\lambda_o^4}{2r^4} , \quad \tanh\Gamma \sim \frac{\pm\lambda_o^2}{r^2} .$$

Thus the only remaining unknown is the gravitational function e^u for which we have three equations (IV.17) - (IV.19). The equations (IV.18) and (IV.19), as in general relativity, can be solved independently. Let us consider the asymptotic form of (IV.18). Using the above approximations and neglecting terms of the order $(\frac{\lambda_o}{r})^5$ and higher we obtain from (IV.18) the result

$$(re^u)' = 1 - \frac{\lambda_o^4}{2r^2 r_o^2} = 1 - \frac{Ge^2}{c^4 r^2} \quad ,$$

which can be integrated as

$$e^u = 1 - \frac{2mG}{c^2 r} + \frac{Ge^2}{c^4 r^2} \quad . \qquad \text{(VII.15)}$$

Now the asymptotic form of the equation (IV.19) is given by

$$(r^2 \; e^u \; u')' = \frac{r_o^2}{r^2} \quad ,$$

which is solved by

$$e^u = 1 - \frac{2mG}{c^2 r} + \frac{G(e^2+g^2)}{c^4 r^2} \quad . \qquad \text{(VII.16)}$$

The two solutions (VII.15) and (VII.16) differ in their last terms. This incompatibility of the two equations (IV.18) and (IV.19) can be resolved by studying the asymptotic form of the remaining equation (IV.17). It is given by

$$\ell_o^2 \; [(\frac{e^u}{r})' + \frac{3}{2r^2} - \frac{r_o^2}{r^4}] = 0 \quad . \qquad \text{(VII.17)}$$

Hence the only solution of (VII.17), which can make the solutions (VII.15) and (VII.16) compatible, is

$$\ell_o^2 = 0 , \qquad \text{(VII.18)}$$

or g = 0. We have thus proved the theorem. The actual value of r where g = 0 and beyond which there is no magnetic charge density involves an indeterminacy. Thus relativistic invariance together with a structure for an elementary particle imposes an indeterminacy on the actual size of the structure. The degree of this indeterminacy may eventually be represented by the introduction of $\hbar$.

VIII. MAGNETIC CHARGE DISTRIBUTION AND THE ABSENCE OF FREE MONOPOLES OF LONG RANGE FIELDS

The properties of the magnetic charge predicted by this theory are novel and bear no relation to other theories on this subject[12,13]. In this theory magnetic charge g does not reside in a magnetically neutral core of an elementary particle as a doublet of positive and negative charges. Furthermore, it does not exist as a free pole carrying positive or negative magnetic charge producing a long range field. Thus the magnetic charge of this theory, besides playing a fundamental role in the creation of mass itself, generates only short range fields associated with strong interactions. In this theory the magnetic charge does not, directly, partake in electromagnetic interactions. The currents of these charges can only give rise to radiation of massive particles instead of the radiation of photons via the long range forces of the electromagnetic field. We have thus, through the magnetic charge, established a classical basis for strong interactions.

From the results of the previous section we see that the "intrinsic magnetic fields" of a particle

$$\mathcal{B}_o = q\, e^{\rho}\cos\Phi\, \sin\theta \; , \; \mathcal{H}_g = \frac{q}{v}\cos\phi \; , \qquad \text{(VIII.1)}$$

for g=0 (i.e. $\Phi=\frac{\pi}{2}$) vanish. The correspondence of the g = 0 solutions to nonregular behavior of the field shows that the intrinsic magnetic fields $\mathcal{B}_o$ and $\mathcal{H}_g$ do not extend beyond the distribution of the neutral and charged magnetic charge densities

$$s^4 = \frac{q}{4\pi}\frac{d}{dr}\left(e^{\rho}\cos\Phi\right)\sin\theta \quad , \qquad \text{(VIII.2)}$$

$$\zeta^4 = \frac{q}{4\pi}\frac{d}{dr}\left(e^{\rho}\cos\Phi\, \sin\Phi\right)\sin\theta$$

respectively, where

$$Q_g^o = \int s^4\, dr d\theta d\phi = \left[q\, e^{\rho}\cos\Phi\right]_o^{\infty} = 0 \; , \qquad \text{(VIII.3)}$$

$$Q_g^{\pm} = \int \zeta^4 \; dr d\theta d\phi = \pm g \quad .$$

However in the asymptotic region $r < \ell_o$ the magnetic field H_g has the form

$$H_g \sim \pm g \frac{r^3 r_o^2}{\ell_o^7} \exp F \quad , \qquad (VIII.4)$$

where F is given by (V.30). It vanishes at the origin r = 0.

For the static distribution, the radial part of the asymptotic equation (II.59) is solved by the spherical Bessel function

$$s^4 = \frac{1}{4\pi} j_1 (\kappa r) \qquad (VIII.5)$$

where the constant κ being large implies that s^4, for $r \neq 0$, is vanishingly small. The zeros of the Bessel function $j_1(\kappa r)$ are given by

$$\tan(\kappa r) = \kappa r \quad . \qquad (VIII.6)$$

At these points the neutral magnetic charge density changes sign and the magnitude of the distribution falls off with alternating signs. Thus at distances where $r >> \ell_o$ the structure of an elementary particle appears to consist of an infinite number of constituent layers of magnetic charge densities. The charge densities of alternating signs are held together by the mutual magnetic and gravitational attractions of the layers.

The absolute value of the total sum of fractional magnetic charges of fixed sign contained in the alternating layers is equal to g. Thus the neutral distribution of magnetic charge in the core of the elementary particle contains the quantity + g of positive magnetic charge and the quantity - g of negative magnetic charge. In general the distribution will depend on the centrifugal magnetic number ℓ (= 0,1,2,...) which is contained in the radial part of the asymptotic equation (II.59) and is associated with the spherical Bessel functions $j_\ell(\kappa r)$. However, the nonlinear equation (II.38) itself may also give rise to a radial magnetic number n (= 0,1,2,...).

In general for the points where the magnetic charge density vanishes one has

$$\frac{q}{4\pi}\frac{d}{dr}\left(e^{\rho}\cos\Phi\right) = 0 \;, \qquad \text{(VIII.7)}$$

or

$$\cos\Phi = \pm\, \ell_o^2\, e^{-\rho} \quad .$$

Now from (VIII.7) we also have

$$\rho' = \Phi' \tan\Phi \quad , \qquad \text{(VIII.8)}$$

where

$$\Phi \neq \frac{\pi}{2} \quad .$$

On using the above in the equation (IV.20) and

substituting for ρ' from (VIII.8) we obtain the non-linear equation

$$\Phi'' \tan\Phi + \frac{3}{2}\frac{\Phi'^2}{\cos^2\Phi} + \frac{v'}{v}\Phi' \tan\Phi - \frac{\lambda_o^4 \Phi'^2 \sin^2\Phi}{\ell_o^4+\lambda_o^4\cos^2\Phi} = 0 \ . \tag{VIII.9}$$

A class of solutions of this equation are given by

$$\Phi(r) = \pm\, n\, \pi \ , \ n=0,1,2,\ldots \tag{VIII.10}$$

and that they are valid at the point r=0. Thus to obtain the remaining solutions we must assume

$$\Phi \neq \pm\, n\, \pi$$

or

$$e^{\rho} \neq \ell_o^2 \ .$$

In this case we can multiply the equation (VIII.9) by $\cot\Phi$ and divide by Φ' to obtain

$$\frac{\Phi''}{\Phi'} + \frac{3}{2}\frac{\Phi'}{\sin\Phi\cos\Phi} + \frac{v'}{v} - \frac{\lambda_o^4 \Phi' \sin\Phi\cos\Phi}{\ell_o^4+\lambda_o^4\cos^2\Phi} = 0 \ . \tag{VIII.11}$$

The equation (VIII.11) is solved by

$$\frac{1}{v} = \frac{\pm 1_o}{2\ell_o}\Phi' (\tan\Phi)^{3/2} \sqrt{(\ell_o^4+\lambda_o^4\cos^2\Phi)} \tag{VIII.12}$$

We thus see that the point of inflexion r=0

(or the point of isolated regularity) and also the indeterminate point r_c corresponding to the case where $\Phi(r_c) = \frac{\pi}{2}$ do not group themselves with the remaining infinite number of interior points $(0 < r < r_c)$ (or interior neutral surfaces) where the neutral magnetic charge density vanishes. Hence we have the restrictions

$$0 < \Phi \leqq \frac{\pi}{2} \quad \text{for} \quad \ell_o^2 \text{ and } \Phi = \cos^{-1}(\ell_o^2 e^{-\rho}) + \frac{\pi}{2} \; ,$$

$$\frac{\pi}{2} < \Phi \leqq \pi \quad \text{for} \; -\ell_o^2 \text{ and } \Phi = \cos^{-1}(-\ell_o^2 e^{-\rho}) - \frac{\pi}{2} \; ,$$

which in terms of e^{ρ} imply, for the points of zero magnetic charge density, the inequality

$$\ell_o^2 \leqq e^{\rho} < \infty \quad . \qquad \text{(VIII.13)}$$

By using the above results and the definition (V.19) for e^{ρ} in the equations (IV.17), (IV.18), (IV.19) (see Appendix 2) they can be integrated at once to yield the solutions

$$e^u = A_1 \frac{t^{3/2}\cosh\Gamma}{\sqrt{(1+t^2)}} + \frac{1}{2}\frac{t^2\cosh^2\Gamma}{\sqrt{(1+t^2)}}\left[\frac{1}{\sqrt{(1+t^2)}} - \frac{K}{2\sqrt{t}}\right] +$$
$$\frac{\kappa^2\ell_o^2}{4}\frac{t^2\cosh^2\Gamma}{\sqrt{(1+t^2)}}\left[\frac{t}{\sqrt{(1+t^2)}} - \frac{3\sqrt{(1+t^2)}}{1+t} + \frac{6E-3K}{2\sqrt{t}} + S(t,\tau_o)\right], \qquad \text{(VIII.14)}$$

$$e^{u} = A_2 \frac{t^{\frac{1}{2}}\cosh^2\Gamma}{\sqrt{(1+t^2)}} + \frac{3t\cosh^2\Gamma}{2\sqrt{(1+t^2)}} \left[\frac{K-2E}{2\sqrt{t}} + \frac{\sqrt{(1+t^2)}}{1+t}\right]$$

$$- \frac{t^2\cosh^2\Gamma}{2(1+t^2)} - \frac{\kappa^2\ell_o^2}{4} \frac{t^3\cosh^2\Gamma}{1+t^2} +$$

$$\frac{5\kappa^2\ell_o^2}{12} \frac{t\cosh^2\Gamma}{\sqrt{(1+t^2)}} \left[\sqrt{(1+t^2)} - \frac{K}{2\sqrt{t}}\right]$$

$$- \frac{\kappa^2\ell_o^2}{4} \frac{t\cosh^2\Gamma}{(1+t^2)} Z(t,\tau_o) , \qquad \text{(VIII.15)}$$

$$\frac{de^u}{dt} + \frac{2te^u}{1+t^2} \tanh^2\Gamma = \frac{A_3 t^{3/2}\cosh^2\Gamma}{(1+t^2)^{3/2}} +$$

$$\frac{\kappa^2\ell_o^2}{4} \frac{t^2\cosh^2\Gamma}{(1+t^2)^{3/2}} \left[\frac{2}{3}\sqrt{(1+t^2)} - \frac{K}{2\sqrt{t}} - Y(t,\tau_o)\right]$$

$$\text{(VIII.16)}$$

In these equations, because of the condition $\Phi \neq \pm n\pi$ the point $t=0$ is excluded. In fact if $t = \frac{r^2}{\ell_o^2}$ then for t large compared to 1 the equations (VIII.15) and (VIII.16) with

$$\tau_o^2 = \frac{e^2}{e^2+g^2}, \quad A_2 = -\frac{2mG}{\ell_o c^2}, \quad A_3 = \frac{2mG}{\ell_o c^2}, \qquad \text{(VIII.17)}$$

yield Schwarzschild solution of general relativity. Hence, as seen from the Appendix 2, the definition (V.19) is correct.

Now, outside $(0<r<r_c)$ the three equations (VIII.14) - (VIII.16) are compatible at only

$r=\infty$ provided $\ell_o=0$. Hence the constant of integration $A_1 = 0$. We thus have three equations to determine the three unknowns t, e^u and $\kappa\ell_o$. By combining (VIII.14) and (VIII.15) one obtains an equation for t which in general would have an infinite number of solutions yielding the surfaces of zero magnetic charge. The equations (VIII.14) - (VIII.16) would further yield, for each surface of zero charge, a relation between A_2, A_3 and ℓ_o, κ. In this way we see that the constant g behaves like an "eigenvalue" of the charge distribution and assumes a spectrum of values. Relativistic invariance of the theory is not compatible with a sharp boundary of charge distribution. Therefore, general covariance of the theory implies, for the surfaces of zero magnetic charge density in the core of an elementary particle, an indeterminacy. The degree of indeterminacy for the surfaces of zero magnetic charge may be given by

$$r\,mc = \hbar$$

or

$$m = \frac{\hbar}{c\ell_o\sqrt{t}}\,, \qquad \text{(VIII.18)}$$

where t is a function of ℓ_o and m. Therefore the solution of the above equations should yield a mass spectrum.[17]

Based on the above results we may now state the fundamental theorem of this theory:

Magnetic Theorem

General relativity and classical electrodynamics are valid only in the region $r > \ell_o$ which corresponds to the g = 0 limit of the generalized theory of gravitation. There exist no free magnetic poles associated with a long range field but elementary particles are composed of stratified layers of neutral magnetic matter with or without electric charge and that the corresponding electric, magnetic and gravitational fields for $g \neq 0$ are regular everywhere. All elementary particles carry an excess of positive and negative magnetic charge which acts as the source of a short range field.

IX. SELF ENERGY AND BINDING ENERGY OF A PARTICLE

We may now use the conserved energy-momentum tensor $\mathfrak{T}^{\mu}_{\nu}$ to calculate the binding energy of a static spherically symmetric system. We shall consider the simplest case of an electrically neutral (i.e. $\lambda_o = 0$) field without polarization charge and, in analogy with general relativity, compute the volume integral of the quantity

$$\mathfrak{T}^4_4 - \mathfrak{T}^1_1 - \mathfrak{T}^2_2 - \mathfrak{T}^3_3 = \frac{q^2}{2\pi} (\sqrt{(-\hat{g})} - \sqrt{(-g)}) \ . \tag{IX.1}$$

By using the definitions (III.16) we may write for the binding energy

$$\Delta E = \int (\mathcal{T}_4^4 - \mathcal{T}_1^1 - \mathcal{T}_2^2 - \mathcal{T}_3^3)\,dr\,d\theta\,d\phi = 2q^2 \int_0^\infty \frac{e^\rho}{v}\,(1-\sin\Phi)\,dr \,. \tag{IX.2}$$

From the field equation (IV.15) we obtain, for $\lambda_o = 0$, the result

$$(v\,e^{u+\rho}u')' = \frac{\kappa^2 e^\rho}{v}\,(1 - \sin\Phi)$$

and hence

$$\Delta E = r_o^2\, q^2\, [v\, e^{u+\rho} u']_o^\infty \,.$$

On substituting from (V.9) and the asymptotic solutions (V.27) - (V.33) and (VII.12) we get the fundamental result

$$\Delta(\pm E) = mc^2 - \frac{(2g_o)^2}{\ell_o} \,, \tag{IX.3}$$

where

$$g_o = 2g\left(\frac{1}{\kappa \ell_o}\right)^3$$

and where $\pm$ signs are due to the linear dependence of energy on v and, as mentioned before, they are interpreted as pertaining to the classical counterparts of particles and anti-particles. The constant of integration m, obtained earlier, is the gravitational mass of the particle. The second term

$$E_S = \frac{(2g_o)^2}{\ell_o} \tag{IX.4}$$

represents the total self energy (or magnetic potential energy) of a totally neutral particle arising from the magnetic attraction between the layers of magnetic charge densities of alternating sign. The factor 2 is due to the two possible signs of a layer of magnetic charge density. Thus E_S may be interpreted as the total rest energy of the constituents of a particle. From (IX.4) we see that for g=0 the self-energy becomes infinite.

For an estimate of the self-energy or the binding energy of an elementary particle we need a reasonable value for the magnetic charge g. One possibility is to assume that the length ℓ_o is of the order of the nucleon Compton wave length (see VIII.19). Another possibility is to relate the gravitational potential energy of a uniform homogeneous spherical nucleon to the dimensionless number $\frac{e^2}{g^2}$ by writing

$$\frac{1}{4\pi}\left(\frac{2}{5}\,\frac{Gm^2}{e^2}\right) = \frac{e^2}{g^2} \quad , \qquad \text{(IX.5)}$$

where if we take m to be the nucleon mass then one of the values of the magnetic charge is given by

$$g = 1 \text{ Coulomb} = 3\times10^9 \text{ e.s.u.}$$

or

$$g = 6.24\times10^{18}\, e \quad . \qquad \text{(IX.6)}$$

The above value for g is, of course, only a guess based on the assumption that the length ℓ_o ought to be of the order of nucleon Compton wave length. The actual value of g may have to come from the quantization of the theory. However, the value (IX.6) appears to be quite reasonable even though, it is 17 orders of magnitude larger than the value of the magnetic charge obtained by Dirac for a free monopole associated with a long range field.

For the value (IX.6) of g the corresponding values of the hadronic (ℓ_o) and leptonic (λ_o) lengths are given by

$$\ell_o \cong \frac{\sqrt{(2G)}}{c^2} g = 1.2\times 10^{-15} \text{ cm.} \qquad (IX.7)$$

$$\lambda_o \cong \frac{\sqrt{(2G)}}{c^2} \sqrt{(eg)} = 4.8\times 10^{-25} \text{ cm.} \qquad (IX.8)$$

The lengths λ_o and ℓ_o as given by (V.12) and (V.13) are equal only if g=e. In this case, as seen from (V.35), the gravitational potential assumes its maximum value. The corresponding value of the length is given by

$$\ell_m = \ell_o = \lambda_o = \frac{2^{3/4}}{c^2} e\sqrt{G} = 2.3\times 10^{-34} \text{ cm.} \qquad (IX.9)$$

A possible speculation on the origin of the length ℓ_m can be based on the assumption that in the "primordial field" the energy density and the corresponding gravitational field were high enough for the particle to consume its binding energy (i.e. its magnetic potential energy) and thus collapse to a size of the order ℓ_m where

$$mc^2 = \frac{e^2+g^2}{r_o} \quad , \qquad \text{(IX.10)}$$

or

$$m = \frac{e}{\sqrt{G}} = 7.2\times10^{-7} \text{ gr.} \qquad \text{(IX.11)}$$

Thus ℓ_m is the smallest size into which particles could have collapsed in the primeval time. The binding energy of such "micro-black holes" and "anti-micro-black holes" is of the order of 10^{15} ergs. However the binding energy of a particle or anti-particle, as follows from (IX.4), is of the order of 10^{33} ergs. Thus the minimum temperature in the primordial field would have been of the order $T_o \sim 10^{50}$ K. Thus T_o is the minimum temperature required to put all the magnetic charge layers together to produce a nucleon. The large size of the self energy confirms the earlier conclusion of the theory that there exist no free monopoles associated with a long range field.

For g = 0, except for the factor ± 1, we obtain the result of general relativity provided that the range of integration in (IX.2) does not extent beyond the size of the material system with Schwarzschild radius $\frac{2mG}{c^2}$. In this case $E = mc^2$ represents the total energy of matter in terms of the energy-momentum tensor of the matter alone.

The above, from the point of view of the

classical field theory, is a solution of the self-energy problem. The presence of negative energy solutions in a classical theory is a pleasant surprise and not necessarilly a vice. The negative energy solutions imply the necessity for a quantum field theoretical formulation of the classical theory.

The above value of the order of 10^{33} c.g.s. units for the binding energy can also be regarded as the degree of conservation of baryon or lepton charge.

X. CONCLUSIONS

From the classical point of view the results of this paper show that the generalized theory of gravitation, which is based on a correspondence principle, lays the foundations for a regular and divergence free electromagnetic and gravitational and short range interactions. In this theory the new idea is the existence of a magnetic charge g which assumes a spectrum of values and that the regularity of the solutions is due to the finite value of g since for g = 0 the spherically symmetric solutions reduce to Nordström's solution of general relativity. An elementary particle has a magnetically neutral core of matter containing a distribution of alternating positive and negative magnetic charge densities over the stratified layers of the core and carries an excess positive or negative magnetic charge which generates a short range field. A novel result is the emergence of an electrically neutral current, in addition to a charged current, and a corresponding neutral field which appears to be a classical version of the vacuum polarization in quantum electrodynamics. From (VI.12) and (VI.13) we see that the ratio of the neutral charge density J_o^4 to the electric charge density in the neighborhood of the origin is given by $(\frac{r_o}{\ell_o})^4$ and becomes infinite for g = 0.

A very interesting consequence of the above results is the finiteness of the self-energy which in turn yields for an elementary particle a finite binding energy. Furthermore, the classical counterparts of the strong and weak interactions are represented by a short range field due to a spectrum of magnetic charge densities superimposed over the neutral magnetic charge distribution in the core of an elementary particle. An important property of the magnetic charge g is its dependence on the mass itself, which is the basic reason for the short range nature of the corresponding field. The appearance of both positive and negative energy solutions with corresponding electric charges is very surprising. In this context it is necessary to study the plane wave solutions of the field equations to see the nature of the negative energy solutions in this case. If $g^+_{\mu\nu}$ represent a set of solutions with positive energy and $g^-_{\mu\nu}$ the corresponding set with negative energy then the superimposed quantities $g^+_{\mu\nu} + g^-_{\mu\nu}$ do not yield an approximate solution though each of them is an exact solution. However, if the fields in $g^{\pm}_{\mu\nu}$ are small compared to q then we may call them weak fields and in this case the field $g^{+}_{\mu\nu} + g^{-}_{\mu\nu}$ is an approximate solution. The approximation can further be improved to higher orders. In particular, one would like to discuss the classical aspects of the scattering of light by light to further understand the relationship between positive and negative energy solutions.

The origin of the negative energy solutions, in this theory, is, presumably due to the presence in the theory of square root terms containing both electromagnetic and gravitational variables. In order to elucidate the negative energy problem we shall calculate the extremum action function of the theory by substituting the field equations (II.33) - (II.35) in the action function (II.26) and obtaining

$$S_{ext.} = - \frac{q^2}{4\pi} \int [\sqrt{(-\hat{g})} - \sqrt{(-g)}] d^4x \ . \qquad (X.1)$$

Hence for each solution of the field equations there exists a nonvanishing extremum action function corresponding to particles represented by the field. The corresponding limit (i.e. $q \to \infty$) of (X.1) yields the extremum value for the action function of general relativity as

$$S_{G\cdot ext.} = - \frac{1}{16\pi c} \int \sqrt{(-g)}\ \Phi^{\mu\nu}\ \Phi_{\mu\nu}\ d^4x \ , \qquad (X.2)$$

which is the action function of a pure electromagnetic field in its own gravitational field.

A Dirac type of linearization (in terms of Dirac matrices) of the action (X.1) was carried out by this author(3) earlier and in view of the, then, unknown physical interpretation and the corresponding solutions of the field equations no useful results were obtained. We are planning to reconsider the older(3) approach in light of this paper's results. The above suggested procedure may turn out to be a simpler way for the quantization of the theory. Quantization is, presumably, the only way to discover a connection between g and e and also to associate negative energy solutions with real particles. However, on a cosmological scale the classical positive and negative energy solutions may imply a large scale symmetry between the distribution of matter and anti-matter in the universe. On a classical level one may regard the total energy content of the universe as being zero. In a similar way total hadron number N_h, total lepton number N_L, total electric charge Q_e and total magnetic charge Q_g of the universe must vanish. In this connection another classical problem that needs early attention is the study of the time dependent spherically symmetric fields.

This paper might have, perhaps, been written over 20 years ago. The mathematical formulation for the theory did not change. However, there are some fundamental reasons for the delay. One of the stumbling blocks in the development of all three versions[1] (Einstein[4], Schrödinger[10]) of the theory has been the misinterpretation of the axial vector s^{μ} (II.44) as the electric current density. We know, now, this was an erroneous assumption. Another important consideration was the absence of a correspondence principle which would, of course, have demonstrated the true nature of s^{μ}. In the affirmative one can cite the recent proliferation of elementary particle models particularly those describing possible constituents of the particles' none of which have been entirely satisfactory. Under these circumstances an elementary particle model based on a fundamental theory, even if a classical one, should be given very serious consideration.

APPENDIX 1

SPHERICALLY SYMMETRIC STATIC AFFINE FIELDS

The static spherically symmetric components of the 64 component affince connections $\Gamma^{\rho}_{\mu\nu}$ can be obtained by a long but straight forward process which leads to an algebraic solution of the equations $\hat{g}_{\mu\nu;\rho} = 0$. For the time independent case there are only 17 nonvanishing components

$$\Gamma^{1}_{11} = -(\tfrac{1}{2}\, u' + \frac{v'}{v}),$$

$$\Gamma^{1}_{22} = \frac{1}{\sin^2\theta}\, \Gamma^{1}_{33} = \tfrac{1}{2}\, v^2 e^{u+\rho}(\Phi'\cos\Phi - \rho'\sin\Phi) \;,$$

$$\Gamma^1_{44} = v^2 e^{2u}\left(\frac{1}{2}u' - 2\Gamma'\tanh\Gamma\right), \quad \Gamma^2_{\{12\}} = \Gamma^3_{\{13\}} = \frac{1}{2}\rho',$$

$$\Gamma^2_{33} = -\sin\theta\cos\theta, \quad \Gamma^3_{\{23\}} = \cot\theta,$$

$$\Gamma^4_{\{14\}} = \frac{1}{2}u' - \Gamma'\tanh\Gamma,$$

$$\Gamma^2_{\{34\}} = \frac{1}{2} v\,\Phi' e^u \tanh\Gamma \sin\theta, \quad \Gamma^3_{\{24\}} = -\frac{1}{2}\frac{v\,e^u}{\sin\theta}\Phi'\tanh\Gamma,$$

$$\Gamma^1_{[23]} = \frac{1}{2} v^2 e^{u+\rho}(\Phi'\sin\Phi + \rho'\cos\Phi)\sin\theta,$$

$$\Gamma^2_{[31]} = -\frac{1}{2}\Phi'\sin\theta, \quad \Gamma^3_{[12]} = -\frac{1}{2}\frac{\Phi'}{\sin\theta},$$

$$\Gamma^1_{[41]} = v\,e^u\,\Gamma', \quad \Gamma^2_{[42]} = \Gamma^3_{[43]} = \frac{1}{2} v\,e^u\,\rho'\tanh\Gamma, \tag{$A_1 1$}$$

where prime indicates differentiation with respect to r.

The only nonvanishing components of $R_{\{\mu\nu\}}$ and $R_{[\mu\nu]}$ are given by

$$R_{11} = \rho'' + \rho'\frac{v'}{v} + \frac{1}{2}(\Phi'^2 + \rho'^2) + \frac{1}{2}\left(u'' + u'^2 + u'\frac{v'}{v} + u'\rho'\right) +$$

$$\Gamma'^2\left(\tanh^2\Gamma - \frac{1}{\cosh^2\Gamma}\right) - \left(\Gamma'' + \frac{v'}{v}\Gamma' + \frac{3}{2}u'\Gamma'\right)\tanh\Gamma. \tag{$A_1 2$}$$

$$R_{22} = -1 + \frac{1}{2} v[v\,e^{u+\rho}(\rho'\sin\Phi - \Phi'\cos\Phi)]'$$

$$- \frac{1}{2}\Phi' v^2 e^{u+\rho}(\Phi'\sin\Phi + \rho'\cos\Phi)$$

$$- \frac{1}{2} v^2 \Gamma' e^{u+\rho}(\rho'\sin\Phi - \Phi'\cos\Phi)\tanh\Gamma = \tag{$A_1 3$}$$

$$-1-\frac{1}{2} v \cosh\Gamma \left[\left(\frac{v\Phi' e^{u+\rho}}{\cosh\Gamma}\right)' \cos\Phi - \left(\frac{v\rho' e^{u+\rho}}{\cosh\Gamma}\right)' \sin\Phi\right]$$

$$R_{33} = R_{22} \sin^2\theta \; , \qquad (A_1 4)$$

$$R_{44}=v^2 e^{2u}[\Gamma'^2-\frac{1}{2}(u''+u'^2+u'\frac{v'}{v}+u'\rho')-\frac{1}{2}(\Phi'^2+\rho'^2)\tanh^2\Gamma+$$

$$\tanh\Gamma \; (2\Gamma'' + 2\Gamma' \frac{v'}{v} + 2\Gamma'\rho' + \frac{3}{2} u'\Gamma')] \; , \qquad (A_1 5)$$

$$\frac{1}{\sin\theta} R_{[23]} = -\frac{1}{2} v[v\, e^{u+\rho}(\Phi'\sin\Phi+\rho'\cos\Phi)]' +$$

$$\frac{1}{2} v^2 e^{u+\rho}[\Phi'(\Phi'\cos\Phi-\rho'\sin\Phi)+\Gamma'\tanh\Gamma(\Phi'\sin\Phi+\rho'\cos\Phi)]=$$

$$(A_1 6)$$

$$-\frac{1}{2} v \cosh\Gamma\left[\left(\frac{v\Phi' e^{u+\rho}}{\cosh\Gamma}\right)' \sin\Phi + \left(\frac{v\rho' e^{u+\rho}}{\cosh\Gamma}\right)' \cos\Phi\right],$$

$$R_{[14]}=-e^{-\rho}[(v\, e^{u+\rho}\rho'\tanh\Gamma)'+\frac{1}{2} v\, e^{u+\rho}(\rho'^2+\Phi'^2)\tanh\Gamma] .$$

$$(A_1 7)$$

Further general relations between Γ's and curvature tensor are given by

$$R_{[\mu\nu]} = -\Gamma^{\rho}_{[\mu\nu]o\rho}$$

$$\Gamma^{\alpha}_{[\mu\nu]} g_{\rho\alpha} + \Gamma^{\alpha}_{[\nu\rho]} g_{\mu\alpha} + \Gamma^{\alpha}_{[\rho\mu]} g_{\nu\alpha} = -\frac{1}{2} I_{\mu\nu\rho}$$

where (o) signifies covariant differentiation with respect to the symmetric part of $\Gamma^{\rho}_{\mu\nu}$ i.e. $\Gamma^{\rho}_{\{\mu\nu\}}$.

APPENDIX 2

OSCILLATIONS OF THE MAGNETIC CHARGE DENSITY

By using the results (VIII.8) and (VIII.12) in the equations (IV.17) - (IV.19) we obtain, for the points of zero magnetic charge density, the equations

$$\left[\frac{(e^{2\rho}-\ell_o^4)^{-\frac{3}{4}}\, e^{\rho} e^{u}}{\cosh^2\Gamma}\right]' = \frac{1}{4}\,\rho'\left[\kappa^2(e^{2\rho}-\ell_o^4)^{\frac{1}{4}}\cosh\Gamma-\kappa^2 e^{-\rho}(e^{2\rho}-\ell_o^4)^{\frac{3}{4}}-2e^{-\rho}(e^{2\rho}-\ell_o^4)^{\frac{1}{4}}\right] \qquad (A_2 1)$$

$$\left[\frac{(e^{2\rho}-\ell_o^4)^{-\frac{1}{4}}\, e^{\rho} e^{u}}{\cosh^2\Gamma}\right]' = \frac{\rho'}{4}(e^{2\rho}-\ell_o^4)^{\frac{1}{4}}\kappa^2\left[e^{\rho}-(e^{2\rho}-\ell_o^4)^{\frac{1}{2}}\cosh\Gamma - \ell_o^4\, e^{-\rho}\right] + \frac{1}{2}\,\rho'\,(e^{2\rho}-\ell_o^4)^{\frac{3}{4}}\, e^{-\rho}\ , \qquad (A_2 2)$$

$$\left[\frac{(e^{2\rho}-\ell_o^4)^{-\frac{1}{4}}}{\cosh^2\Gamma}\left(\frac{u'}{\rho'}+2\tanh^2\Gamma\right)e^{u+\rho}\right]' = \frac{\kappa^2\rho'}{4}(e^{2\rho}-\ell_o^4)^{\frac{1}{4}}\left[e^{\rho}-\frac{(e^{2\rho}-\ell_o^4)^{\frac{1}{2}}}{\cosh\Gamma}\right]\ . \qquad (A_2 3)$$

On dividing by ρ' and using the substitution

$$e^{\rho} = \ell_o^2\sqrt{(1+t^2)}\ ,\quad \cosh\Gamma = \frac{\sqrt{(1+\frac{e^2}{g^2}+t^2)}}{\sqrt{(1+t^2)}}\ ,$$

$$\tanh\Gamma = \frac{\pm e}{g} \frac{t}{\sqrt{(1+\frac{e^2}{g^2}+t^2)}} ,$$

the above equations can be replaced by

$$\frac{d}{dt}\left[\frac{\sqrt{(1+t^2)}}{t^{3/2}\cosh^2\Gamma} e^u\right]$$

$$= \frac{1}{4}\frac{t^{3/2}}{1+t^2}\left[\kappa^2\ell_o^2\left(\cosh\Gamma - \frac{t}{\sqrt{(1+t^2)}}\right) - \frac{2}{\sqrt{(1+t^2)}}\right] , \quad (A_2 4)$$

$$\frac{d}{dt}\left[\frac{\sqrt{(1+t^2)}}{t^{1/2}\cosh^2\Gamma} e^u\right]$$

$$= -\frac{1}{4}\frac{t^{5/2}}{1+t^2}\left[\kappa^2\ell_o^2\left(\cosh\Gamma - \frac{t}{\sqrt{(1+t^2)}}\right) - \frac{2}{\sqrt{(1+t^2)}}\right] , \quad (A_2 5)$$

$$\frac{d}{dt}\left[\frac{(1+t^2)^{3/2}}{t^{3/2}\cosh^2\Gamma}\left(\frac{de^u}{dt} + \frac{2te^u}{1+t^2}\tanh^2\Gamma\right)\right]$$

$$= \frac{1}{4}\kappa^2\ell_o^2\frac{t^{3/2}}{1+t^2}\left[\sqrt{(1+t^2)} - \frac{t}{\cosh\Gamma}\right] . \quad (A_2 6)$$

The integrals for the integration of these equations are given below:

$$\text{(i)} \int \frac{t^{3/2}}{1+t^2}\, dt = 2t^{1/2} + \frac{1}{\sqrt{2}}\left[-\tan^{-1}\left(\frac{\sqrt{(2t)}}{1-t}\right) + \tanh^{-1}\left(\frac{\sqrt{(2t)}}{1+t}\right)\right] ,$$

$$\text{(ii)} \int \frac{t^{5/2}}{(1+t^2)^{3/2}}\, dt = -\frac{t^{3/2}}{\sqrt{(1+t^2)}} + \frac{3}{2}\int \frac{\sqrt{(t)}}{\sqrt{(1+t^2)}}\, dt ,$$

where

$$\int \frac{\sqrt{(t)}}{\sqrt{(1+t^2)}}\, dt = K(\alpha,\tfrac{1}{\sqrt{2}})-2E(\alpha,\tfrac{1}{\sqrt{2}})+2\,\frac{\sqrt{t}\sqrt{(1+t^2)}}{1+t} \quad .$$

The first and second kind elliptic integrals are defined by

$$K(\alpha,\ \tfrac{1}{\sqrt{2}}) = \int^{\alpha} \frac{d\gamma}{\sqrt{(1-\frac{1}{2}\sin^2\gamma)}} \quad ,$$

$$E(\alpha,\ \tfrac{1}{\sqrt{2}}) = \int^{\alpha} \sqrt{(1-\tfrac{1}{2}\sin^2\gamma)}\ d\gamma \quad ,$$

respectively, where

$$\alpha = \cos^{-1}\left(\frac{1-t}{1+t}\right) \ , \ \frac{dK}{dt} = \frac{1}{\sqrt{\ [t(1+t^2)]}}, \ \frac{dE}{dt} = \frac{\sqrt{(1+t^2)}}{\sqrt{(t)}(1+t)^2} \quad ,$$

$$\text{(iii)} \int \frac{t^{3/2}}{(1+t^2)^{3/2}}\, dt = -\frac{\sqrt{t}}{\sqrt{(1+t^2)}} + \frac{1}{2}\,K(\alpha,\ \tfrac{1}{\sqrt{2}}) \quad ,$$

$$\text{(iv)} \int \frac{t^{3/2}}{\sqrt{(1+t^2)}}\, dt = \frac{2}{3}\left[\sqrt{t}\sqrt{(1+t^2)}-\frac{1}{2}\,K(\alpha,\ \tfrac{1}{\sqrt{2}})\right] \quad ,$$

$$\text{(v)} \int \frac{t^{5/2}}{t^2+1}\, dt$$

$$= \frac{2}{3}\,t^{3/2} - \frac{1}{\sqrt{2}}\left[\tanh^{-1}\left(\frac{\sqrt{(2t)}}{1+t}\right) + \tan^{-1}\left(\frac{\sqrt{(2t)}}{1-t}\right)\right] \quad ,$$

(vi) $\displaystyle\int \frac{t^{7/2}}{(1+t^2)^{3/2}}\,dt$

$$= -\frac{t^{5/2}}{\sqrt{(1+t^2)}} + \frac{5}{3}\left[\sqrt{t}\,\sqrt{(1+t^2)} - \frac{1}{2}K\left(\alpha, \frac{1}{\sqrt{2}}\right)\right] .$$

(vii) $\displaystyle\int \frac{t^{3/2}}{1+t^2}\cosh\Gamma\, dt = \sqrt{t}\; S(t,\tau_o)$,

(viii) $\displaystyle\int \frac{t^{5/2}}{1+t^2}\cosh\Gamma\, dt = \sqrt{t}\; Z(t,\tau_o)$,

(ix) $\displaystyle\int \frac{t^{5/2}}{1+t^2}\frac{dt}{\cosh\Gamma} = \sqrt{t}\; Y(t,\tau_o)$.

The functions S, Z and Y can be expanded in powers of $\frac{e^2}{g^2}$. We shall retain only the term proportional to $\frac{e^2}{g^2}$. Thus

$$S(t,\tau_o) = \frac{1}{\sqrt{t}}\int \frac{t^{3/2}}{1+t^2}\cosh\Gamma\, dt$$

$$= \frac{1}{\sqrt{t}}\int \frac{t^{3/2}}{1+t^2}\left[1 + \frac{e^2}{g^2}\,\frac{1}{(1+t^2)}\right]^{\frac{1}{2}} dt$$

$$\sim \frac{1}{\sqrt{t}}\int \frac{t^{3/2}}{1+t^2}\left[1 + \frac{e^2}{2g^2}\,\frac{1}{(1+t^2)}\right] dt$$

$$= 2 - \frac{e^2}{4g^2}\,\frac{1}{(1+t^2)} + \frac{1}{\sqrt{(2t)}} \times$$

$$\left[\left(1+\frac{e^2}{8g^2}\right)\tanh^{-1}\left(\frac{\sqrt{(2t)}}{1+t}\right) - \left(1-\frac{e^2}{8g^2}\right)\tan^{-1}\left(\frac{\sqrt{(2t)}}{1-t}\right)\right],$$

where

$$\int \frac{t^{3/2}}{(1+t^2)^2}\, dt = - \frac{\sqrt{t}}{2(1+t^2)} + \frac{1}{4}\int \frac{t^{-\frac{1}{2}}}{1+t^2}\, dt \ ,$$

$$\int \frac{t^{-\frac{1}{2}}}{1+t^2}\, dt = \frac{1}{\sqrt{2}} \left[\tanh^{-1}\left(\frac{\sqrt{(2t)}}{1+t}\right) + \tan^{-1}\left(\frac{\sqrt{(2t)}}{1-t}\right)\right] \ .$$

For the Z-function we have

$$Z(t,\tau_o) = \frac{1}{\sqrt{t}} \int \frac{t^{5/2}}{1+t^2} \cosh\Gamma \, dt$$

$$\sim \frac{1}{\sqrt{t}} \int \frac{t^{5/2}}{1+t^2} \left[1 + \frac{e^2}{2g^2} \frac{1}{(1+t^2)}\right] dt$$

$$= \frac{2t}{3} - \frac{e^2}{4g^2} \frac{t}{1+t^2} - \frac{1}{\sqrt{(2t)}} \times$$

$$\left[\left(1-\frac{3e^2}{8g^2}\right)\tan^{-1}\left(\frac{\sqrt{(2t)}}{1-t}\right) + \left(1+\frac{3e^2}{8g^2}\right)\tanh^{-1}\left(\frac{\sqrt{(2t)}}{1+t}\right)\right] ,$$

where

$$\int \frac{\sqrt{t}}{1+t^2}\, dt = \frac{1}{\sqrt{2}} \left[\tan^{-1}\left(\frac{\sqrt{(2t)}}{1-t}\right) - \tanh^{-1}\left(\frac{\sqrt{(2t)}}{1+t}\right)\right] \ .$$

The function $Y(t,\tau_o)$ can be obtained from the function Z by replacing $\frac{e^2}{g^2}$ by $-\frac{e^2}{g^2}$. Thus

$$Y(t,\tau_o) = \frac{2t}{3} + \frac{e^2}{4g^2}\frac{t}{1+t^2} - \frac{1}{\sqrt{(2t)}} \times$$

$$\left[\left(1+\frac{3e^2}{8g^2}\right)\tan^{-1}\left(\frac{\sqrt{(2t)}}{1-t}\right)+\left(1-\frac{3e^2}{8g^2}\right)\tanh^{-1}\left(\frac{\sqrt{2t}}{1+t}\right)\right].$$

Finally we note that all of the above computations could also have been carried out in terms of the function $\Phi(r)$ where

$$t = \tan\Phi \quad .$$

For t large compared to 1 the elliptic functions K and E tend to $-\frac{2}{\sqrt{t}}$.

APPENDIX 3

BASIC IDENTITIES OF THE NONSYMMETRIC FIELD

Under an arbitrary transformation of the coordinates by

$$x'^{\mu} = f^{\mu}(x)$$

we have the transformation rules

$$\hat{g}'_{\mu\nu} = \frac{\partial x^{\rho}}{\partial x'^{\mu}}\frac{\partial x^{\sigma}}{\partial x'^{\nu}}\hat{g}_{\rho\sigma} \quad , \qquad (A_3 1)$$

and, as follows from the infinitesimal parallel displacement law

$$\delta A^{\mu} = -\Gamma^{\mu}_{\rho\sigma}A^{\rho}\delta x^{\sigma} \quad ,$$

the rules

$$\Gamma'^{\rho}_{\mu\nu} = \frac{\partial x^{\alpha}}{\partial x'^{\mu}} \frac{\partial x^{\beta}}{\partial x'^{\nu}} \frac{\partial x'^{\rho}}{\partial x^{\sigma}} \Gamma^{\sigma}_{\alpha\beta} - \frac{\partial^2 x'^{\rho}}{\partial x^{\alpha} \partial x^{\beta}} \frac{\partial x^{\alpha}}{\partial x'^{\mu}} \frac{\partial x^{\beta}}{\partial x'^{\nu}} \quad . \tag{$A_3 2$}$$

From ($A_3 1$) we obtain

$$\surd(-\hat{g}') = \left|\frac{\partial x}{\partial x'}\right| \surd(-\hat{g}) \quad , \tag{$A_3 3$}$$

where

$$\left|\frac{\partial x}{\partial x'}\right| = \mathrm{Det}\left[\frac{\partial x^{\mu}}{\partial x'^{\nu}}\right] \quad .$$

We also have

$$\hat{g}'^{\mu\nu} = \left|\frac{\partial x}{\partial x'}\right| \frac{\partial x'^{\mu}}{\partial x^{\rho}} \frac{\partial x'^{\nu}}{\partial x^{\sigma}} \hat{g}^{\rho\sigma} \quad . \tag{$A_3 4$}$$

We may now apply an infitesimal transformation of the coordinates

$$x'^{\mu} = x^{\mu} + \xi^{\mu}(x) \quad , \tag{$A_3 5$}$$

and obtain

$$\left|\frac{\partial x'}{\partial x}\right| \cong 1 + \xi^{\rho}_{,\rho} \quad , \quad \left|\frac{\partial x}{\partial x'}\right| \cong 1 - \xi^{\rho}_{,\rho} \quad , \tag{$A_3 6$}$$

where ξ^{μ} are small compared to 1. Hence

$$\delta \hat{g}^{\mu\nu} = \hat{g}'^{\mu\nu}(x) - \hat{g}^{\mu\nu}(x) = \hat{g}^{\mu\rho}\xi^{\nu}_{,\rho} + \hat{g}^{\rho\nu}\xi^{\mu}_{,\rho} - \hat{g}^{\mu\nu}\xi^{\rho}_{,\rho} - \hat{g}^{\mu\nu}_{,\rho}\xi^{\rho} , \tag{$A_3 7$}$$

$$\delta\Gamma^{\rho}_{\mu\nu} = \Gamma^{\alpha}_{\mu\nu}\xi^{\rho}{}_{,\alpha}-\Gamma^{\rho}_{\mu\alpha}\xi^{\alpha}{}_{,\nu}-\Gamma^{\rho}_{\alpha\nu}\xi^{\alpha}{}_{,\mu}-\Gamma^{\rho}_{\mu\nu,\alpha}\xi^{\alpha}-\xi^{\rho}{}_{,\mu\nu} \ . \qquad (A_3 8)$$

From the definition

$$R_{\mu\nu} = -\ \Gamma^{\rho}_{\mu\nu,\rho}+\Gamma^{\rho}_{\mu\rho,\nu}+\Gamma^{\rho}_{\mu\sigma}\Gamma^{\sigma}_{\rho\nu}-\Gamma^{\rho}_{\mu\nu}\Gamma^{\sigma}_{\rho\sigma} \ ,$$

and from using $(A_3 8)$ we obtain

$$\delta R_{\mu\nu} = -\ R_{\mu\nu,\rho}\xi^{\rho}-R_{\rho\nu}\xi^{\rho}{}_{,\mu}-R_{\mu\rho}\xi^{\rho}{}_{,\nu} \ , \qquad (A_3 9)$$

which follows also from

$$R'_{\mu\nu}(x+\xi)= \frac{\partial x^{\rho}}{\partial x'^{\mu}}\,\frac{\partial x^{\sigma}}{\partial x'^{\nu}}\,R_{\rho\sigma} = (\delta^{\rho}_{\mu}-\xi^{\rho}{}_{\mu})(\delta^{\sigma}_{\nu}-\xi^{\sigma}{}_{,\nu})R_{\rho\sigma} \ ,$$

or

$$R'_{\mu\nu}(x) + R'_{\mu\nu,\rho}\xi^{\rho}\cong R_{\mu\nu}(x)- R_{\mu\sigma}\xi^{\sigma}{}_{,\nu} - R_{\rho\nu}\xi^{\rho}{}_{,\mu} \ .$$

Hence

$$\delta R_{\mu\nu}=R'_{\mu\nu}(x)-R_{\mu\nu}(x)\cong-R_{\mu\nu,\rho}\xi^{\rho}-R_{\mu\sigma}\xi^{\sigma}{}_{,\nu}-R_{\rho\nu}\xi^{\rho}{}_{,\mu} \ .$$

The above results lead to

$$R_{\mu\nu}\ \delta\hat{g}^{\mu\nu} = \hat{g}^{\mu\nu}(R_{\underset{+-}{\mu\nu;\rho}}-R_{\underset{++}{\mu\rho;\nu}}-R_{\underset{--}{\rho\nu;\mu}})\xi^{\rho} \ , \qquad (A_3 10)$$

$$\hat{g}^{\mu\nu}\ \delta R_{\mu\nu} = \hat{g}^{\mu\nu}_{+-;\rho}\ \delta\Gamma^{\rho}_{\mu\nu} - \hat{g}^{\mu\nu}_{+-;\nu}\ \delta\Gamma^{\rho}_{\mu\rho} \ , \qquad (A_3 11)$$

where the divergence terms have been dropped and where we assumed that the variations $\delta\hat{g}^{\mu\nu}$, $\delta\Gamma^{\rho}_{\mu\nu}$

arise from the infinitesimal change of the co-ordinates as given by ($A_3 5$). We further assumed that

$$\hat{g}^{\mu\nu}_{\underset{+-}{}\,;\rho} = 0 \quad .$$

The changes, under ($A_3 5$), in $\sqrt{(-\hat{g})}$, $\sqrt{(-g)}$ are given by (dropping divergence terms)

$$\delta\sqrt{(-\hat{g})} = \frac{1}{2}\,\hat{g}_{\mu\nu}\;\delta\hat{g}^{\mu\nu} = 0 \quad , \qquad (A_3 12)$$

$$\delta\sqrt{(-g)} = \frac{1}{2}\,b_{\mu\nu}\;\delta\hat{g}^{\mu\nu} = 0 \quad .$$

These results can be used in

$$\delta\sqrt{(-\hat{g})} = \frac{1}{2}\,\hat{g}_{\mu\nu}(\hat{g}^{\mu\rho}\xi^{\nu}{}_{,\rho} + \hat{g}^{\rho\nu}\xi^{\mu}{}_{,\rho} - \hat{g}^{\mu\nu}\xi^{\rho}{}_{,\rho} - \hat{g}^{\mu\nu}{}_{,\rho}\xi^{\rho})$$

$$= \frac{1}{2}\,\hat{g}^{\mu\nu}(\hat{g}_{\underset{+-}{\mu\nu};\rho} - g_{\underset{++}{\mu\rho};\nu} - \hat{g}_{\underset{--}{\rho\nu};\mu})\xi^{\rho} \qquad (A_3 13)$$

and in the variational principle $\delta\, S = 0$, to derive the results

$$\hat{g}^{\mu\nu}(R_{\underset{+-}{\mu\nu};\rho} - R_{\underset{++}{\mu\rho};\nu} - R_{\underset{--}{\rho\nu};\mu}) = 0 \quad , \qquad (A_3 14)$$

$$\hat{g}^{\mu\nu}(\hat{g}_{\underset{+-}{\mu\nu};\rho} - \hat{g}_{\underset{++}{\mu\rho};\nu} - \hat{g}_{\underset{--}{\rho\nu};\mu}) = 0 \quad . \qquad (A_3 15)$$

The differential identities ($A_3 14$) can also be derived from the definition and the various symmetries of the curvature tensor

$$R^{\sigma}{}_{\mu\nu\rho} = -\Gamma^{\sigma}_{\mu\nu,\rho}+\Gamma^{\sigma}_{\mu\rho,\nu}+\Gamma^{\sigma}_{\alpha\nu}\Gamma^{\alpha}_{\mu\rho}-\Gamma^{\alpha}_{\mu\nu}\Gamma^{\sigma}_{\alpha\rho} . \quad (A_3 16)$$

From the forms (II.51) and (II.52) of $(A_3 14)$ and $(A_3 15)$ it is clear that we may add to $g_{\mu\nu}$ (or $R_{\{\mu\nu\}}$) a constant multiple of $b_{\mu\nu}$ and also we may add to $\Phi_{\mu\nu}$ (or $R_{[\mu\nu]}$) a constant multiple of $F_{\mu\nu}$ and the identities remain unchanged. Hence the identities $(A_3 14)$ and $(A_3 15)$ are also satisfied by the tensor $A_{\mu\nu} = \hat{g}_{\mu\nu} + \lambda(b_{\mu\nu}+F_{\mu\nu})$. In fact by direct substitution of $A_{\mu\nu}$ in $(A_3 15)$ and carrying out indicated covariant differentiations lead to

$$\hat{g}^{\mu\nu}(b_{\mu\nu,\rho}-b_{\mu\rho,\nu}-b_{\rho\nu,\mu}+2b_{\rho\sigma}\Gamma^{\sigma}_{\mu\nu}+F_{\mu\nu,\rho}+F_{\nu\rho,\mu}+F_{\rho\mu,\nu})=$$

$$\surd(-g)b^{\mu\nu}(b_{\mu\nu,\rho}-2b_{\mu\rho,\nu})+2b_{\rho\sigma}\,\hat{g}^{\mu\nu}\Gamma^{\sigma}_{\mu\nu} = 0 ,$$

where we used the relations

$$\frac{1}{2}\, b^{\mu\nu}b_{\mu\nu,\rho} = \frac{[\surd(-g)]_{,\rho}}{\surd(-g)} , \quad \hat{g}^{\mu\nu}\Gamma^{\rho}_{\mu\nu} = -\,[\surd(-g)b^{\rho\nu}]_{,\nu} . \quad (A_3 17)$$

We may now use $(A_3 9)$ to calculate the conservation laws of the field. Thus we may write

$$\hat{g}^{\mu\nu}\delta R_{\mu\nu}=[(\hat{g}^{\mu\nu}R_{\mu\rho}+\hat{g}^{\nu\mu}R_{\rho\mu}-\delta^{\nu}_{\rho}\hat{g}^{\alpha\beta}R_{\alpha\beta})_{,\nu}+\hat{g}^{\alpha\beta}{}_{,\rho}R_{\alpha\beta}]\xi^{\rho}, \quad (A_3 18)$$

where the divergence term has been dropped. In order to calculate the last term in ($A_3$18) we shall calculate the variation of $\hat{g}^{\mu\nu}{}_{,\rho}$. From the definition

$$\hat{g}^{\mu\nu}{}_{,\rho} = -\hat{g}^{\alpha\nu}\Gamma^{\mu}_{\alpha\rho} - \hat{g}^{\mu\alpha}\Gamma^{\nu}_{\rho\alpha} + \hat{g}^{\mu\nu}\Gamma^{\alpha}_{\rho\alpha} ,$$

we obtain

$$\delta\hat{g}^{\mu\nu}{}_{,\rho} = \Gamma^{\sigma}_{\rho\sigma}\delta\hat{g}^{\mu\nu} - \Gamma^{\mu}_{\sigma\rho}\delta\hat{g}^{\sigma\nu} - \Gamma^{\nu}_{\rho\sigma}\delta\hat{g}^{\mu\sigma} +$$

$$\hat{g}^{\mu\nu}\delta\Gamma^{\sigma}_{\rho\sigma} - \hat{g}^{\mu\sigma}\delta\Gamma^{\nu}_{\rho\sigma} - \hat{g}^{\sigma\nu}\delta\Gamma^{\mu}_{\sigma\rho} .$$

On multiplying through by

$$\beta^{\rho}_{\mu\nu} = \delta^{\rho}_{\mu}\Gamma^{\alpha}_{\nu\alpha} - \Gamma^{\rho}_{\mu\nu} , \qquad (A_3 19)$$

we obtain

$$\beta^{\rho}_{\mu\nu}\,\delta\hat{g}^{\mu\nu}{}_{,\rho} = \hat{g}^{\mu\nu}\delta\beta_{\mu\nu} + 2\beta_{\mu\nu}\,\delta\hat{g}^{\mu\nu} , \qquad (A_3 20)$$

where

$$\beta_{\mu\nu} = \Gamma^{\rho}_{\mu\sigma}\Gamma^{\sigma}_{\rho\nu} - \Gamma^{\rho}_{\mu\nu}\Gamma^{\sigma}_{\rho\sigma} , \ \beta = \hat{g}^{\mu\nu}\beta_{\mu\nu} . \qquad (A_3 21)$$

Now on using the variation of β we get the result

$$\delta\beta = \beta_{\mu\nu}\delta\hat{g}^{\mu\nu} + \hat{g}^{\mu\nu}\delta\beta_{\mu\nu} = -\beta_{\mu\nu}\delta\hat{g}^{\mu\nu} + \beta^{\rho}_{\mu\nu}\delta\hat{g}^{\mu\nu}{}_{,\rho} .$$

Hence the variational derivative of β yields the

result

$$\frac{\partial \beta}{\partial \hat{g}^{\mu\nu}} - \frac{\partial}{\partial x^\rho}\left(\frac{\partial \beta}{\partial \hat{g}^{\mu\nu}{}_{,\rho}}\right) = - R_{\mu\nu} \quad . \qquad (A_3 22)$$

We may now use $(A_3 22)$ to obtain

$$- \hat{g}^{\mu\nu}{}_{,\rho} R_{\mu\nu} = \frac{\partial}{\partial x^\sigma}\, [\delta^\sigma_\rho \beta - \hat{g}^{\mu\nu}{}_{,\rho} \beta^\sigma_{\mu\nu}] \quad . \qquad (A_3 23)$$

On substituting the result $(A_3 23)$ in $(A_3 18)$ we obtain the conservation law

$$\mathfrak{T}^\nu_{\mu,\nu} = 0 \quad , \qquad (A_3 24)$$

where

$$-4\pi\kappa^2 q^{-2} \mathfrak{T}^\nu_\mu = \hat{g}^{\nu\rho} R_{\mu\rho} + \hat{g}^{\rho\nu} R_{\rho\mu} - \delta^\nu_\mu \hat{g}^{\rho\sigma} R_{\rho\sigma} + \hat{g}^{\rho\sigma}{}_{,\mu} \beta^\nu_{\rho\sigma} - \delta^\nu_\mu \beta \quad . \qquad (A_3 25)$$

By using the field equations (II.33) and (II.34) in $(A_3 25)$ we may rewrite it as

$$-4\pi \mathfrak{T}^\nu_\mu = q^2 \delta^\nu_\mu [\surd(-\hat{g}) - \surd(-g) - \tfrac{1}{2} \hat{g}^{[\rho\sigma]} F_{\rho\sigma}] + q^2 \hat{g}^{[\rho\nu]} F_{\rho\mu} + q^2 \kappa^{-2} [\hat{g}^{\rho\sigma}{}_{,\mu} \beta^\nu_{\rho\sigma} - \delta^\nu_\mu \beta] \quad , \qquad (A_3 26)$$

or

$$\mathfrak{T}^\nu_\mu = \frac{q^2}{4\pi} [\delta^\nu_\mu (\surd(-\hat{g}) - \surd(-g)) - \hat{g}^{[\nu\rho]} F_{\mu\rho}] + \tfrac{1}{2} \delta^\nu_\mu \mathfrak{T} - \frac{q^2 \kappa^{-2}}{4\pi} \hat{g}^{\rho\sigma}{}_{,\mu} \beta^\nu_{\rho\sigma} , \qquad (A_3 27)$$

where

$$\mathfrak{T} = \mathfrak{T}^\mu_\mu \quad , \quad \hat{g}^{\rho\sigma}{}_{,\mu} \beta^\mu_{\rho\sigma} = 2\beta \quad .$$

REFERENCES

1. B. Kursunoglu, Phys. Rev. 88, 1369 (1952).
2. B. Kursunoglu, Rev. Mod. Phys. 29, 412 (1957).
3. B. Kursunoglu, Nuovo Cimento 15, Series X, 729 (1960).
4. Einstein and Schrödinger versions of the generalized theory of gravitation,(5,6,7,8) because of the absence of a correspondence principle in them, do not yield equations (II.57). In fact the Einstein and Schrödinger theories are obtained from the present one by setting $r_o = \infty$ (!). Hence these theories cannot yield Lorentz's equations of motion.
5. A. Einstein, Can. J. Math. 2, 120 (1950).
6. B. Kaufman, Helv. Physica Acta Supp. 4, 227 (1956).
7. A. Einstein and B. Kaufman, Ann. Math. 62, 128 (1955).
8. A. Papapetrou, Proc. Roy. Irish Acad. Vol. LII, Sec. A, No. 6, 69 (1948).
9. The presence of a short range neutral charge density J_o^4 may be thought of as the classical version of the vacuum polarization in quantum electrodynamics.
10. In this theory the correspondence with general relativity plus Maxwell's equations is based on setting a physical constant like magnetic charge g equal to zero while in quantum theory correspondence with classical mechanics is obtained by setting $\hbar = 0$. However, if a

relation between the g of this theory and h can be established then in this theory also the correspondence principle can be satisfied by setting $\hbar = 0$ everywhere.

11. The nonconservation of the neutral charge density for $g = 0$ can be compared to the divergence of the vacuum polarization in quantum electrodynamics.

12. P.A.M. Dirac, Phys. Rev. 74, 817 (1948).

13. J. Schwinger, Proceedings of Coral Gables Conference, Symmetry Principles at High Energy, (Freeman, San Francisco, 1966) Eds. A. Perlmutter, et al.

14. E. Schrödinger, Proc. Roy. Irish Acad. LI, A213 (1948).

15. The functions Φ and Γ are expressible in terms of Ω and Λ and are therefore invariant under coordinate transformations.

16. Because of (III.18), (IV.3) and also in view of the spectrum of values assumed by the magnetic charge, the radius R remains invariant under coordinate transformations.

17. We observe that because of the invariance of the function $\Phi(r)$ and the relation $t=\tan\Phi$ all the statements concerning the core of an elementary particle are in complete agreement with the principle of general covariance.

THE GAUGE THEORETIC BREAKTHROUGH?*†

M. A. B. Bég

The Rockefeller University

New York, New York 10021

I. INTRODUCTION

After several years of relative tranquility, Particle Physics is once again in a state of considerable agitation. The excitement centers around the subject of spontaneously broken gauge theories,[1] a subject which has its origins in the seminal contributions of Weinberg,[2] in 1967, and Salam,[3] in 1968. These authors revived the old notion that weak interactions and electromagnetism have a common origin in a (non-abelian) gauge principle[4] but made a break with the past in suggesting that the full gauge symmetry be broken spontaneously,[5] a la Higgs and

---

*Invited Report to the Coral Gables Conference, January 8, 1974.

†Work supported in part by the U. S. Atomic Energy Commission under Contracts AT(11-1)-2232.

Kibble, to the level of electromagnetic gauge-invariance. The hope was that the so-called Higgs mechanism would permit the theory to retain enough memory of the gauge group to stay renormalizable - in contrast to the situation in which one breaks the symmetry manifestly by explicit insertion of mass terms in the Lagrangian. This hope found fulfillment[6] in the work of 't Hooft, 't Hooft and Veltman, B. W. Lee and Lee and Zinn-Justin. This development was followed by a period of intense activity in the model building industry.[7] Among the weak interaction models which were constructed, some met specific theoretical desiderata, some attempted to follow the shifting winds of experimental trends and some had no particular raison d'etre at all. It soon became apparent that while it was easy to construct models if one permitted oneself the freedom to postulate arbitrary numbers of quarks (and leptons), things got tight if one attempted to assign the quarks to specific roles in hadron dynamics. The problem of a gauge invariant genesis of broken hadron symmetries emerged - and still remains - as the most challenging problem facing theoretical physicsts. The prospect of a quick solution to the problem of weak interactions slowly faded away as one began to get bogged down in the intricacies of strong interactions. The discovery by Politzer and by Gross and Wilczek that unbroken gauge theories based on semi-simple Lie Groups are asymptotically free,[8] while kindling fresh hopes for a gauge-theoretic synthesis of weak, electromagnetic and strong interactions, has raised a host of new problems; the net effect has been a moratorium on model construction while people strive for a deeper understanding of the structure of gauge theories.

In this talk I shall present, in language as non-technical as I can muster, a somewhat selective review of some of these developments. To set the stage for a serious consideration of gauge theories, I shall first review the canonical difficulties of weak interaction theory.

II. THE STATE OF WEAK INTERACTIONS[9]

1. Notation, etc.

In order to facilitate the discussion, I shall assume that all hadrons are "built up" out of a set of fundamental spin $\frac{1}{2}$ objects. Whatever be the quantum numbers assigned to these subnuclear objects, I shall refer to them as quarks.[10] For dealing with low-lying hadrons it will be sufficient to use a quark-triplet, (p,n,λ), which carries the fundamental representation of SU(3). The (p,n) constitute an iso-doublet with hypercharge $+\frac{1}{3}$, the λ is an iso-singlet with hypercharge $-\frac{2}{3}$. I shall use the same symbols, p and n, to denote the proton and neutron states; reference to context will minimize the possibility of confusion.

I should stress that the use of quarks in the present context, as a convenient prop for making manifest certain group-theoretic properties, is without prejudice to the question of real existence or non-existence of these particles.

2. The Universal Fermi Interactions

All weak interactions, that are known to exist at currently available energies, appear to be adequately described (except for CP-violating effects) by the phenomenological current x current interaction:

$$H_{wk} = \frac{G_F}{\sqrt{2}} \cdot 4\, J_\mu^{(+)\dagger}\, J_\mu^{(+)}, \tag{1}$$

with

$$\begin{aligned} J_\mu^{(+)} &= \bar{p}\, \gamma_\mu \left(\frac{1-\gamma_5}{2}\right) (n \cos\theta + \lambda \sin\theta) \\ &+ \bar{\nu}_e\, \gamma_\mu \left(\frac{1-\gamma_5}{2}\right) e^- + \bar{\nu}_\mu \gamma_\mu \left(\frac{1-\gamma_5}{2}\right) \mu^- . \end{aligned} \tag{2}$$

The present numerical values for the Fermi constant[11] G_F and the Cabibbo angle[12] θ are:

$$G_F = (1.4350 \pm 0.0002) \times 10^{-49} \text{ ergs. cm}^3$$
$$\cong (1.023 \pm 0.001) \times 10^{-5}/M_p^2, \tag{3}$$

$$\sin\theta \cong 0.22 - 0.24. \tag{4}$$

Eqs. (1) and (2) make manifest a bare universality between the weak interactions of quarks and leptons; a bridge between this bare universality and the physical universality as reflected, for example, in the near equality of coupling constants in various β-decays and μ-decay, is established via CVC,[13] current algebra,[14] and PCAC.[15]

We note that in the conventional picture of weak interactions outlined above, neutral (i.e. $\Delta Q=0$) currents are conspicuous by their absence. The best experimental limits are on $\Delta Q = 0$, $|\Delta S| = 1$ currents and arise from the process $K_L^o \to \mu^+\mu^-$; the present value of the rate[16]

$$\Gamma(K_L \to \mu^+\mu^-) = 1.1 \times 10^{-8}\, \Gamma(K_L \to \text{all}), \tag{5}$$

is consistent with the hypothesis that the process proceeds entirely through the sequence[17]

$K_L^o \to 2\gamma \to \mu^+ \mu^-$. In this connection, it is worth noting that one cannot attribute the near absence of $K_L^o \to \mu^+ \mu^-$ to a cancellation between $\Delta S = +1$ and $\Delta S = -1$ amplitudes; the simultaneous presence of currents with both signs of ΔS would give rise to $|\Delta S| = 2$ amplitudes in leading order. This would lead to an unacceptably large $K_L - K_S$ mass difference and also run afoul of limits established in decay processes:[12]

$$\frac{\Gamma(\Xi^o \to p\pi^-)}{\Gamma(\Xi^o \to \Lambda\pi^o)} < 3.5 \times 10^{-5} \ (90\% C.L.). \qquad (6)$$

3. Difficulties of the Fermi Theory

The above picture of weak interactions is satisfactory only on a superficial level; the moment one tries to probe deeper, by using the H_{wk} of Eq. (1) to calculate anything other than the low energy amplitudes it is contrived to reproduce, one runs into serious difficulties.

Consider first the problem of (electromagnetic) radiative corrections.[9,18] Calculation of these corrections is mandatory since effects of order $G_F\alpha$ are within the realm of measurability. In the leptonic sector one finds that (a) the corrections to μ-decay are finite to all orders in α but (b) the corrections to $\nu_e + e^- \to \nu_e + e^-$ diverge in order α^2. In the semi-leptonic sector one finds that if one ignores strong interaction effects, the corrections to $n \to p + e^- + \bar{\nu}_e$ diverge already in order α; the hope that this divergence might be tamed by the damping effect of strong interactions was dashed to the ground by the work of Dicus, Norton, Abers and Quinn.[19] In the presence of the electromagnetic field, therefore, the observed pattern of weak interactions cannot be said to follow from Eqs. (1) and (2).

Next, let us consider the extrapolation of Eq. (1) to the high-energy domain. As is well known, the amplitudes for $\nu_\mu + e^- \to \mu^- + \nu_e$ and $\nu_e + e^- \to \nu_e + e^-$, in the Fermi theory, overtake their unitarity bounds at about 300 GeV and 450 GeV, respectively. (These processes are pure S-wave, in the Fermi theory, and the ratio of the predicted cross section to the unitarity bound rises as the fourth power of the c.m. energy!). If one tries to avoid this "unitarity castrophe" by looking upon the H_{wk} of Eq. (1) as a field theoretic interaction that might conceivably yield a unitary S-matrix when one liberates oneself from the Born approximation, one gets into more serious troubles. All higher order effects of H_{wk} are infinite[20] in an unrenormalizable way and one ends up losing control even over the low-energy phenomenology.

The situation is quite untenable; all the difficulties stem, of course, from the bad high energy behaviour of H_{wk} which is also responsible for its lack of renormalizability (in the conventional sense).

4. The Intermediate Vector Boson (IVB) Theory

In this formulation one starts with the Yukawa interaction

$$H_I = \frac{g}{\sqrt{2}} J_\mu^{(+)} W^{+\mu} + \text{Herm. Conj.} \tag{7}$$

Where W^+ is the field operator for positively charged vector bosons and $J_\mu^{(+)}$ is the current in Eq. (2). It is evident from the covariant form of the W-propagator

$$D_F^{\mu\nu}(k) = i \frac{(k^\mu k^\nu / m_W^2) - g^{\mu\nu}}{k^2 - m_W^2} \tag{8}$$

that so long as one restricts oneself to transitions in which the momentum transfer and the masses of the participating particles are small compared to m_W, the second order effects of H_I in Eq. (7) can be simulated by the H_{wk} of Eq. (1) provided one makes the identification

$$\frac{g^2}{8m_W^2} = \frac{G_F}{\sqrt{2}} \tag{9}$$

Our discussion of low energy phenomenology within the framework of the Fermi theory carries over mutatis mutandis to the Intermediate Boson or Yukawa theory. The high energy behaviour of the Yukawa interaction is, of course, very different from that of the Fermi interaction; unfortunately, however, in the present context, it is not more benign.

5. Difficulties of the IVB Theory

The difficulties faced by the IVB theory are different in detail from the difficulties faced by the Fermi theory; however, the overall pattern of problems is not significantly altered.

In the matter of electromagnetic radiative corrections, one finds now that in most processes the radiative corrections diverge in all orders of α. The clash with unitarity, in order g^2, does not look too severe if one restricts one's attention to processes such as $\nu_\mu + e^- \to \mu^- + \nu_e$ or $\nu_e + e^- \to \nu_e + e^-$; the partial wave amplitudes in these processes (they are not pure S-wave in the IVB theory) violate their unitarity bounds only logarithmically with the energy. This logarithmic overtake of the unitarity bounds, by the Born approximation, occurs in theories as respectable as quantum electrodynamics; however, in QED one knows how to generate a unitary S-matrix, by including higher order effects, and here one does not. In any case the more spectacular clash with unitarity, with cross sections violating their unitarity bounds according to a power of the energy, that was seen in the Fermi theory, is visible also in the IVB theory if one looks at other, well chosen, processes; for example, in the process $\nu_\mu + \bar{\nu}_\mu \to W^+ + W^-$ the helicity amplitude $(-\frac{1}{2},\frac{1}{2}) \to (0,0)$ overtakes the unitarity bound at[21] $\sim 10^3$ GeV.

As is the case with the Fermi theory, the difficulties of the IVB theory stem from its bad high energy behavior which in turn is responsible for its lack of renormalizability. However, quite unlike the situation with the Fermi theory, there is now a faint light at the end of the tunnel. The culprit in the IVB theory can be identified as the $k_\mu k_\nu$ term in the W-propagator; if some means could be found for taming this term one would have the beginnings of a sensible theory of weak interactions.

We note that an example of a theory in which the offending term in the vector propagator causes no problems is afforced by massive electrodynamics. In this theory, however, not only does the vector meson couple to a conserved current but there is a latent gauge freedom as well; this freedom can be used to quantize the theory in a gauge in which it is manifestly renormalizable.[22]

6. Attempts to Extract Finite Answers

There have been many attempts to get finite answers out of non-renormalizable theories; these are invariably based on the hope that the difficulties of such theories stem from the unjustified use of perturbation theory, that the theory will somehow damp itself if one can liberate oneself from the perturbative framework. The basis for such hope has its genesis in many-body theory.[23] In the context of weak interactions, however, efforts to lend substance to this hope have so far proved futile.

Before we give serious consideration to a theory we must have some assurance that it is more than a statement of theological views; in particular, it must permit calculations and yield a reasonable number of experimentally testable predictions. The limitations imposed by the present state of calculational technology oblige us, therefore, to stay - at least for the time being - within the framework of renormalizable theories. Next, let us assume that the current x current form of the interaction is not a total red herring and that V-A currents do have a natural place in the theory; in other words, let us reject the "deception" alternatives discussed in the literature.[24] This narrows the choice down to gauge theories, and to these I now turn.

III. THE GAUGE ALTERNATIVE

The formulation of renormalizable theories which unify weak and electromagnetic interactions rests upon a combination of subtle theoretical ideas and elegant theorems. The first idea is the demand that the theory be invariant under local gauge transformations. This introduces in a natural manner the vector mesons which will be ultimately identified with the photon and the intermediate vector mesons of the weak interactions. The general problem of constructing theories which are invariant under a local group G was solved some time ago[25] by Yang and Mills and Utiyama.

An important theoretical development, of fairly recent origin, in the study of Yang-Mills theories has been the derivation of a consistent set of Feynman rules.[26] A method that has proved to be very fruitful in this connection is the path integral quantization. An important consequence of this investigation is that the Feynman rules for non-Abelian gauge theories are not simply given by the vertices of the original Lagrangian, but additional vertices and propagators associated with the fictitious complex scalar fields obeying Fermi statistics must also be included. These fields are generally referred to as Faddeev-Popov ghosts.

While a theory of weak interactions based

on an unbroken gauge principle is a renormalizable theory, the masslessness of gauge fields such as $W^{\pm}$ precludes the possibility of any contact with physical reality; one must introduce masses for all gauge fields save the photon, in such a way that the theory retains enough memory of gauge invariance to stay renormalizable. The second important idea, which ultimately solves this problem, is spontaneous symmetry breaking; namely, the vacuum state or ground state is not invariant under the full symmetry of the Lagrangian.

If a symmetry is broken only by the asymmetry of the vacuum state, the Goldstone theorem[27] would indicate that there is a massless spin zero excitation - a Goldstone boson - corresponding to each generator which fails to annihilate the vacuum. A theory afflicted with such massless excitations is clearly unacceptable, on experimental grounds. However, the proof of the Goldstone theorem is contingent on the simultaneous use of (a) positivity of the metric in Hilbert space and (b) manifest covariance or locality. In a gauge theory both (a) and (b) cannot be simultaneously valid. [If we quantize the theory in the Coulomb gauge, (a) is valid but (b) is not; in a covariant gauge the situation reverses.] The possibility of evading the Goldstone theorem in the context of a gauge theory was first noted by Higgs[5] in an abelian model; Higgs' work was generalized to the non-abelian case by Kibble.[5] What happens in a spontaneously broken gauge theory is quite remarkable: the dual plagues of massless gauge quanta and unwanted Goldstone bosons simply neutralize each other. The would-be Goldstone bosons coalesce with the Yang-Mills quanta, providing the longitudinal degree of freedom needed to generate massive vector bosons. (Vector mesons, such as the photon, which couple to non-trivial generators that annihilate the vacuum remain massless.) Furthermore, the renormalizability of the original massless Yang-Mills theory is preserved inspite of the spontaneous symmetry breaking! The proof of this remarkable fact is rather subtle and was achieved only recently.[6]

These serendipitous discoveries have led to the view that the gauge strategy, implemented a la Higgs and Kibble, may hold the key to the riddle of weak interactions. An attractive, perhaps compelling, feature of this approach is the achievement of a very natural and painless synthesis of weak and electromagnetic interactions. Indeed, the way may have been opened for a grand synthesis of <u>all</u> elementary particle interactions.

IV. MODELS

1. Basic Rules and General Features

One can abstract, from the work of Weinberg,[2] Salam[3] and others,[1] some general rules for the construction of gauge-theoretic models of weak and electromagnetic interactions. These may be summarized as follows:

(a) Choose a gauge group G.

(b) Assign the left and right chiral projections of the spin $\frac{1}{2}$ fields to representations of G. The assignment should be such as to maintain weak interaction phenomenology and cancel the triangle anomalies which interfere with renormalizability.[28,29]

(c) Introduce spin zero (Higgs) fields with suitable transformation properties under G.

(d) Introduce gauge fields via

$$\partial_\mu \to \partial_\mu - i\Sigma_a\, g_a \{\vec{t}\cdot\vec{A}_\mu\}_a \qquad (10)$$

where the summation is over the simple pieces of the algebra of G and the $\vec{t}$ are the matrix representatives of the group generators.

(e) Couple the Higgs fields, ϕ, invariantly and renormalizably to themselves in such a way that the potential energy has a local maximum at $\phi = 0$.

(f) Introduce G-invariant Yukawa couplings of the Higgs fields to the spin $\frac{1}{2}$ fields.

(g) Trigger the Higgs-Kibble mechanism via $\phi = \langle 0|\phi|0\rangle + \phi'$.

It is evident that one can construct infinitely many gauge models almost all of which would appear, a priori, to be on an equal footing. Before embarking on model building, therefore, it is necessary to enunciate some principles of economy or elegance and lay down the specific desiderata that the model is supposed to meet.

To motivate some of the specific models that have been proposed in the literature, and to provide a framework for discussing these models and classifying them, let us consider some general features that are, of necessity, present in all gauge models:

(a) $G \supset SU(2)$. This follows from the fact that any Yukawa theory of weak interactions must contain at least two mediating fields W^+ and W^-; the sources of these fields are the weak charges Q_{Wk} and Q^+_{Wk} which must belong to the Lie algebra of G. Since Q_{Wk} cannot commute with Q^+_{Wk} there is a third non-trivial generator

$$Q_3 = \frac{1}{2}\,[Q_{Wk}, Q^+_{Wk}], \tag{11}$$

which must also belong to the Lie algebra of G.

(b) Neutral Weak Currents or/and Heavy Leptons. The existence of the electrically neutral generator Q_3 automatically implies the existence of neutral weak currents, except if Q_3 is proportional to Q, the electric charge; in this exceptional case there need not be any neutral currents other than the electromagnetic current. For $Q_3 \sim Q$ the commutation relation (11) cannot be realized in the leptonic Hilbert space if ν_μ, μ^-, ν_e, e^- are the only available fields. Additional leptons must be introduced to implement Eq. (11); these are presumably fairly massive since they have not been observed as yet. Hence the oxymoron: Heavy Leptons.

(c) Charm. Since Q_{Wk} contains both $\Delta S = 0$ and $\Delta S = 1$ pieces, it is evident that Q_3 will, in general, contain both $\Delta S = +1$ and $\Delta S = -1$ pieces and thereby lead to severe difficulties; without additional input, one will end up with unacceptably large amplitudes for $K_L \rightarrow \mu^+ \mu^-$ and $\Delta S = 2$ transitions in the hadronic sector. The most popular way of resolving this problem follows the suggestion of Maiani, Illiopoulos and Glashow.[30] One enlarges the hadronic Hilbert space to include particles carrying an additive quantum number, felicitously called charm, which is conserved by all save the weak interactions. If charmed particles are sufficiently massive, one can augment Q_{Wk} with a charm changing piece without disturbing the phenomenology of weak interactions at low energies and in such a way as to exorcise the $\Delta S \neq 0$ pieces in Q_3. Thus with $n_c = n \cos\theta + \lambda \sin\theta$, $\lambda_c = -n \sin\theta + \lambda \cos\theta$, and

$$Q_{Wk} = \int d^3x \, \{p^+ n_c + p'^+ \lambda_c\}, \tag{12}$$

we have

$$Q_3 = \frac{1}{2} \int d^3x \, \{p^+ p + p'^+ p' - n^+ n - \lambda^+ \lambda\} \tag{13}$$

which is free of $\Delta S \neq 0$ pieces.

The qualitative enlargement of the hadron spectrum into the charmed sector is an important element in the gauge-theoretic approach; it leads to a cross-coupling between the problems of weak interactions and hadron spectroscopy and furnishes a useful point of contact with experiment.

It is often convenient to label models by the nature of their charm content.

2. Existing Models; An Overview

Apart from some exceptions the hitherto published models have - quite commendably - tended to economize on the size of the gauge group; most restrict themselves to groups no larger than U(1) ⊗ SU(2), the group of the Weinberg-Salam model.

The first set of models which followed the W-S model, the Lee-Prentki-Zumino (L-P-Z) model[31] and the Georgi-Glashow (G-G) model[32] - to cite the best known examples - focused primarily on the nature of the neutral current. In the L-P-Z model the weak neutral current is rendered neutrino-free by the introduction of two charged heavy leptons; the model was in fact designed primarily to provide an avenue of escape for the gauge strategy, in case neutral current experiments made things uncomfortable for the W-S model. The G-G model goes a step further; by introducing two charged and two neutral heavy leptons, the weak neutral current is eliminated completely. Since the only gauge fields in the G-G model are $W^{\pm}$ and γ, the model has a measure of aesthetic appeal to it. While recent experiments appear to indicate that neutral currents involving ν_{μ}'s indeed exist[33] the dust has yet to settle; in the meantime the L-P-Z and G-G models cannot be ruled out on purely objective grounds.

The second set of models were designed to meet the needs of hadron spectroscopy; tailored, rather, to implement the gauge-theoretic philosophy within the framework of the conventional wisdom on hadron structure. In this class are the "quartet charm" models[28,34] and the "triplet charm" models.[35] None of these models solve the problem of broken hadron symmetries; however, by taking cognizance of hadron symmetries they illuminate some aspects of the problem and constitute natural stepping stones to more satisfactory theories.

Along with models which attempt to address themselves to problems of a broad and "global" nature, there exist a whole host of models constructed primarily for the purpose of investigating very specific and localized problems. These are too numerous to mention here, and I refer you to the literature.[1]

Finally, note that despite the fact that the gauge-theoretic approach is plagued by a number of very profound, unsolved, theoretical problems, despite the fact that experiment has given no convincing indication that we are on the right track, the moratorium on model-construction is not universally observed. One particular sphere in which I anticipate many contributions in the near future is the class of models that I choose to describe as Super Models.[36] Super Models are models which attempt to synthesize all elementary particle interactions at one blow, putting leptons and baryons into the same pot (i.e., the same irreducible representation of the gauge group). Neither lepton conservation nor baryon conservation is exact in such models, the only conserved quantities being total fermion number and electric charge; however, the extent to which these selection rules are violated is adjustable and the parameters in the theory can be adjusted to meet the present lower limit on, say, the proton lifetime.

3. Experimental Status of Gauge Models

As of today, no one has seen intermediate vector bosons, heavy leptons or charmed hadrons. However there is mounting experimental evidence in favor of neutral currents. If the work of the CERN group[33] is confirmed by other experimenters, and survives the test of time, it would be a signal triumph for the gauge approach. On a more mundane level, the neutral current data suffices to rule out models in which neutral currents are neutrinoless in leading order. The status of a couple of surviving models may be gleaned from Table I.

V. UNSOLVED PROBLEMS

A theory of weak interactions can be deemed to be satisfactory if it meets the following requirements:

(a) It should reproduce the known phenomenology in leading order.

(b) It should be within the framework of an acceptable model of hadron structure.

(c) The symmetry structure of weak interactions, manifest in leading order, should be stable under inclusion of higher order effects. Universality [μ - e, lepton-hardon and hadron-hadron (e.g., π^+ - C^{12})] should be maintained, modulo small tolerable corrections. Second order weak effects should be genuine second order weak effects, not uncontrollably large, of order $G_F\alpha$, say. Sacrosanct selection rules such as no semi-leptonic $\Delta Q = 0$, $\Delta S = 1$ transitions, no non-leptonic $\Delta S = 2$, etc. should be violated within acceptable limits. Similar remarks apply to other weak interaction selection rules that might be built into the theory. For example, if the theory implements the (non-leptonic) $\Delta I = \frac{1}{2}$ rule on a kinematical Lagrangian level, it should also ensure that $\Delta I \neq \frac{1}{2}$ amplitudes emerging in higher orders have the observed magnitudes.

(d) It should not do much violence to symmetries respected by the strong interactions. Parity and strangeness violation effects should be $O(G_F)$, not $O(\alpha)$; isotopic spin breakage should be $O(\alpha)$ not $O(1)$.

(e) The theory should permit <u>broken</u> SU(3) and $SU(2)_L \otimes SU(2)_R$ on the Lagrangian level (and - hopefully - permit the S-matrix to retain some memory of these symmetries).

(f) The theory should be strictly renormalizable[37] (i.e., without introduction of new counter terms) with the desiderata (c) and (d) fulfilled in a natural, rather than an artificial way. In other words, we demand that the symmetry breakage effects be finite and calculable and of the correct magnitude; the magnitude should not have to be adjusted by hand after renormalization. I share the view that this requirement is more than just a matter of aesthetic taste.

Having stated these criteria I can mention the first unsolved problem: no theory known at the moment meets them. The demand most difficult to meet is that of complete naturality (i.e., naturality for all the symmetries in the game). The "domain of naturality" of some theories may be enlarged by enlarging the gauge group.[38] Large gauge groups and the concomitant plethora of Higgs fields have their own drawbacks; these may be overlooked, however, if significant progress is achieved.

A theory would be compelling if it meets the requirements (a) - (f) and, in the process, also sheds some light on other long-standing problems of weak-interaction theory. In other words, the "true" theory may be expected to fulfill some additional desiderata:

(g) It should illuminate the nature of μ - e universality by providing some explanation for the empirical relationship

$$\frac{m_e}{m_\mu} = \frac{3\alpha}{\pi} \ln 2.$$

(h) It should provide some insight into the origins of the Cabibbo angle.

(i) It should furnish an explanation for CP-violation, in particular for the remarkable restriction of CP-violating effects to the K_L - K_S mass-matrix.

(j) It should yield a deeper understanding of nature's abhorrence for strangeness-changing neutral currents.

It should perhaps be obvious that the major unsolved problems of weak interaction theory lie in the hadronic domain, and stem from the need to cater to hadron symmetries. While it has long been known that the structure of weak interactions is intertwined with the symmetries of strong interactions, the demands of the strong on the weak were never quite so formidable as in the gauge context. Extensions of the gauge strategy to strong interactions per se, and attempts at a grand synthesis of all elementary particle interactions have lent a new dimension to the search for a solution.

VI. CONCLUDING REMARKS

I would like to conclude with a brief assessment of future prospects.

The resolution of the problem of weak-interactions appears to be inextricably intertwined with the solution of what might have appeared at first sight to be a deeper problem:

the synthesis of strong, electromagnetic and weak interactions in a unified gauge theoretic framework. The problems that stand in the way of such a synthesis are formidable indeed.
However, the rewards are so great that the theoretical efforts must, and undoubtedly will, continue. One should not forget that "formidable problems" may well be "great opportunities".

The elusiveness of a satisfactory model, despite two to three years of intensive investigation by a variety of physicists, strongly suggests the importance of searching for new phenomenological clues. It seems to me that the breakthrough will probably come after experimental effort has extended the frontiers of phenomenology, and narrowed the scope of theoretical speculation in a decisive way.

Quantity	Quartet Charm Model	Triplet Charm Model: Purely Elastic	Triplet Charm Model: Purely Inelastic	Experiment
$[\sigma(\bar{\nu}_e+e^-\to\bar{\nu}_e+e^-)-\sigma_o]/\sigma_o$	-0.09	-0.25 to 1.2	$0(\alpha)$	0.1±0.8
$\sigma(\nu_\mu+e^-\to\nu_\mu+e^-)/\sigma_o$	0.21	<0.48 (Input)	$0(\alpha^2)$	<0.48
$\sigma(\bar{\nu}_\mu+e^-\to\bar{\nu}_\mu+e^-)/\sigma_o$	0.31	R_2	$0(\alpha^2)$	<1.63
$\dfrac{[\sigma(\nu_\mu+p\to\nu_\mu+\pi^o+p)+\sigma(\nu_\mu+n\to\nu_\mu+\pi^o+n)]}{2\sigma(\nu_\mu+n\to\mu^-+\pi^o+p)}$	>0.37	<0.09	$0(\alpha^2)$	- (Free N)
	>0.22	<0.05	$0(\alpha^2)$	<0.14 (N in Al)
$\dfrac{\sigma(\nu_\mu+N\to\nu_\mu+\ldots.)}{\sigma(\nu_\mu+N\to\mu^-+\ldots.)}$	0.28±0.014	<(0.10±0.02)	$0(\alpha^2)(E<E_c)$	0.23±0.03
			$>0.014(E>>E_c)$	
$\dfrac{\sigma(\nu_\mu+N\to\nu_\mu+\ldots.)}{\sigma(\bar{\nu}_\mu+N\to\bar{\nu}_\mu+\ldots.)}$	1.8±0.18	1	$1(E>>E_c)$	1.25±0.29
m_W	67 GeV	14 to 32GeV	18 GeV	?

TABLE I
PRESENT STATUS OF SOME GAUGE MODELS

σ_o is the cross section for $\bar{\nu}_e + e^- \to \bar{\nu}_e + e^-$ in the Feynman-Gell-Mann theory, E_c is the charm threshold. All experimental numbers are from Mussets' report in the Proceedings of the Aix-en-Provence Conference (1973). For a detailed description of the models and the phenomenological analysis see Refs. (1) and (39).

REFERENCES

1. For a recent review see, M. A. B. Bég and A. Sirlin, "Gauge Theories of Weak Interactions", Rockefeller University Report No. COO-2232B-47 (1974). (To be published in the Annual Reviews of Nuclear Science, Vol. 24.)
2. S. Weinberg, Phys. Rev. Letters 19, 1264 (1967).
3. A. Salam, Proceedings of the Eighth Nobel Symposium (John Wiley, N. Y., 1968).
4. J. Schwinger, Ann. Phys. (N.Y.) 2, 407 (1957); S. L. Glashow, Nucl. Phys. 22, 579 (1961); A. Salam and J. C. Ward, Physics Letters 13, 168 (1964).
5. P. W. Higgs, Phys. Letters 12, 132 (1964); Phys. Rev. Letters 13, 508 (1964); Phys. Rev. 145, 1156 (1966). See also F. Englert and R. Brout, Phys. Rev. Letters 13, 321 (1964); G. S. Guralnik, C. R. Hagen and T. W. B. Kibble, Phys. Rev. Letters 13, 585 (1964); T. W. B. Kibble, Phys. Rev. 155, 1554 (1967).
6. G. 't Hooft, Nucl. Phys. B33, 173 (1971); ibid, B25, 167 (1971); G. 't Hooft and M. Veltman, Nucl. Phys. B50, 318 (1972); B. W. Lee, Phys. Rev. D5, 823 (1972); B. W. Lee and J. Zinn-Justin, Phys. Rev. D5, 3121, 3137 and 3155 (1973); ibid, D7, 1049 (1973).

7. For a review of the situation as of September 1972 see: B. W. Lee, Proceedings of XVI Int. Conf. on High Energy Physics, NAL (1972).

8. H. D. Politzer, Phys. Rev. Letters 30, 1346 (1973); D. Gross and F. Wilczek, Phys. Rev. Letters 30, 1343 (1973).

9. See also the presentation of these topics in: M. A. B. Bég, Lectures delivered at the 1970 Brookhaven Summer School, BNL Report No. 15732 (1970) p. 321.

10. M. Gell-Mann, Phys. Rev. 125, 1067 (1962) and Phys. Letters 8, 214 (1964).

11. A. M. Sachs and A. Sirlin, "Muon Decay", Contribution to "Muon Physics", edited by V. H. Hughes and C. S. Wu (to be published).

12. See e.g. B. Stech, Proc. of the II Int. Conf. on Elementary Particle Physics, Aix-en-Provence (1973).

13. R. P. Feynman and M. Gell-Mann, Phys. Rev. 109, 193 (1958); S. Gerstein and J. Zeldovich, Zh. Ersperim. i Teor. Fiz. 29, 698 (1955). [Translation: JETP 2, 575 (1956)].

14. M. Gell-Mann, Physics 1, 63 (1964).

15. J. Bernstein, S. Fubini, M. Gell-Mann and W. Thirring, Nuovo Cim. 17, 757 (1960); Y. Nambu, Phys. Rev. Letters 4, 380 (1960); Chuo Kuang-Chao, JETP 12, 492 (1961).

16. W. C. Carrithers et. al., Phys. Rev. Letters 30, 1336 (1973).
17. M. A. B. Bég, Phys. Rev. 132, 426 (1963); M. L. Sehgal, Nuovo Cim. 45, 785 (1966) and Phys. Rev. 183, 1511 (1969).
18. A. Sirlin, Acta Physica Austriaca Suppl. V, 353 (1968) and Proceedings of the Topical Conference on Weak Interactions, CERN, Geneva, p. 409 (1969).
19. E. Abers, D. Dicus and R. Norton, Phys. Rev. Letters 18, 676 (1967). E. Abers, D. Dicus, R. Norton and H. Quinn, Phys. Rev. 167, 1461 (1968).
20. c.f. W. Heisenberg, Z. fur Phys. 101, 533 (1936); Ann. der Physik 32, 20 (1938).
21. F. E. Low, Comments in Nuc. and Part. Phys. 2, 33 (1968).
22. W. Zimmerman, Springer Tracts in Modern Physics 50, 143 (1969).
23. T. D. Lee, Nuovo Cim. 59A, 579 (1969).
24. See e.g. M. Gell-Mann, M. L. Goldberger, N. M. Kroll and F. E. Low, Phys. Rev. 179, 1518 (1969).
25. C. N. Yang and R. L. Mills, Phys. Rev. 96, 191 (1954); R. Utiyama, Phys. Rev. 101, 1597 (1956).
26. R. P. Feynman, Acta Phys. Polonica 24, 697 (1963); B. S. DeWitt, Phys. Rev. Letters 12, 742 (1964); Phys. Rev. 162, 1195 and 1239 (1967); S. Mandelstam, Phys. Rev. 175, 1580, 1604 (1968); L. D. Faddeev and

V. N. Popov, Phys. Letters 25B, 29 (1967); L. D. Faddeev, Theor. and Math. Phys. 1, 3 (1969) [English translation: Theor. and Math. Phys. 1, 1 (1970)]. E. S. Fradkin and I. V. Tyutin, Phys. Rev. D2, 2841 (1970); M. Veltman, Nuc. Phys. B21, 288 (1970); R. L. Mills, Phys. Rev. D3, 2960 (1971); R. N. Mohapatra, Phys. Rev. D4, 378, 1007, 2215 (1971); ibid, D5, 417 (1972); S. Weinberg, Phys. Rev. D7, 1068 (1973) and L. F. Li, ibid, 3815 (1973).

27. J. Goldstone, A. Salam and S. Weinberg, Phys. Rev. 127, 965 (1962).

28. C. Bouchiat, J. Iliopoulos and Ph. Meyer, Phys. Letters 38B, 519 (1972).

29. D. Gross and R. Jackiw, Phys. Rev. D6, 477 (1972); H. Georgi and S. L. Glashow, Phys. Rev. D6, 429 (1972).

30. S. L. Glashow, J. Iliopoulos and L. Maiani, Phys. Rev. D2, 1285 (1970).

31. B. W. Lee, Phys. Rev. D6, 1188 (1972); J. Prentki and B. Zumino, Nucl. Phys. B47, 99 (1972).

32. H. Georgi and S. L. Glashow, Phys. Rev. Letters 28, 1494 (1972).

33. F. J. Hasert et. al., Phys. Lett. 46B, 138 (1973) and Nuc. Phys. B. (to be published).

34. J. C. Pati and A. Salam, Phys. Rev. D8, 1240 (1973).

35. M. A. B. Bég and A. Zee, Phys. Rev. Letters 30, 675 (1973).
36. J. C. Pati and A. Salam, Phys. Rev. Letters 31, 661 (1973); H. Georgi and S. L. Glashow, Phys. Rev. Letters 32, 438 (1974).
37. K. Symanzik, in "Fundamental Interactions at High Energies II", eds. A. Perlmutter, R. W. Williams and G. J. Iverson (Gordon and Breech, N. Y., (1972)).
38. See e.g. M. A. B. Bég, Phys. Rev. D8, 664 (1973).
39. M. A. B. Bég, Phys. Letters (in press).

INSTABILITIES OF MATTER IN STRONG EXTERNAL FIELDS AND AT HIGH DENSITY†*

Abraham Klein and Johann Rafelski

Department of Physics

University of Pennsylvania

Philadelphia, Pennsylvania 19174

ABSTRACT

In the recent literature, there have been a number of theoretical studies of new phenomena involving the instability of the ground state (sometimes called the vacuum state) of various fundamental forms of matter under unusual perturbations. In this paper we consider in detail two examples of instability under the application of Coulomb fields of superheavy nuclei, against the emission of positrons and the creation of a condensate of charged pion pairs. The former may occur for

---

†Supported in part by the U.S. Atomic Energy Commission.

*For presentation at Orbis Scientiae, Center for Theoretical Studies of the University of Miami, Jan. 7-12, 1974.

atomic number $Z \sim 170$, whereas the latter requires $Z > 10^3$. The theory of possible forms of pion condensation inside nuclear matter and neutron star matter is also discussed. By careful attention to fundamentals, an effort is made to clarify some of the controversy now raging in the literature concerning these speculations.

I. INTRODUCTION

The purpose of this report is to review some recent theoretical developments which predict instabilities in the behavior of various forms of matter when subjected to sufficiently strong external electrostatic fields or when compressed to sufficiently high density. Except for the common feature of instability these phenomena, in detail, are quite distinct, as we shall see by a specific discussion of several examples. The best established theoretically(1-11) is that an assembly of localized positive charges confined to a region of nuclear size ("superheavy nucleus") becomes unstable against the emission of a pair of positrons as we increase the nuclear charge number, Z, beyond $Z_c \sim 170$. Further discrete thresholds occur (the next one at $Z \sim 185$) until ultimately ($Z > 10^3$) we would see positive muon pairs. Since $Z_c \geq 170$ will soon be available, albeit transiently, in heavy ion collisions, the above phenomena may be the first of those discussed to be observed. In section II, the basic theoretical physics needed to understand this instability will be described. The relatively minor but interesting change in the structure of the nuclear charge distribution is seen to be a consequence of Fermi-Dirac statistics.

By contrast, we turn next to an academic example,[12,9] sufficiently instructive in our view, however, that we devote sections III and IV to its study. In this example, we imagine that charged pi mesons ($\pi_{\pm}$) are subject only to electrostatic interactions. We then show that the same (stripped) nuclear charge distribution described above would, for sufficiently large $Z > 10^3$, be unstable against the formation of an overall neutral cloud of $\pi_{\pm}$ pairs. The system can be stabilized by including the self-Coulomb interaction of the pion cloud. Remarkably enough, the condensation energy of these pions might ultimately be sufficient to neutralize the Coulomb repulsion of the protons. It is seen in these discussions how the contrasting phenomena are related to the character of the solutions of the single-particle Dirac and Klein-Gordon equations in an external potential.

In the final two sections we turn to the potentially most interesting, but also, at the moment, the most controversial of the phenomena reviewed in this paper. Various authors, principally Migdal in the USSR[11-16], Sawyer, Scalapino, and collaborators in the U.S.[17-19] and more recently Baym and Flowers[20,21] and others[22,23] have suggested and made model calculations illustrating the possible occurrence (or predicting the non-occurrence) of several kinds of pion condensation phenomena in nuclear matter (N, number of neutrons = Z, number of protons) or inside neutron stars ($N >> Z$). These include neutral pion and $\pi_{\pm}$ pair condensates in nuclear matter, π_- condensates in neutron stars, and π_o and $\pi_{\pm}$ pair condensates in neutron stars. The basic mechanism throughout is the attractive p-wave pion-nucleon interaction.

In our discussion, we have sought to extract some stable elements from the conflicts and the contrasting methodology. Thus, in section V, we have confined ourselves to purely formal results, which in our view establish a method for the study of these possible phase transitions. Within this context we are able to clarify some of the conflicts in methodology.

It seems to us that none of the calculations made so far is truly cogent, but several are highly suggestive. In section VI, we emphasize that perhaps the most accessible portion of the problem is the search for the transition point for a second order phase change. This problem is quite closely tied to the challenge of deriving from first principles the optical potential for pion-nucleus collisions. We conclude with some remarks concerning this relationship.

II. SOLUTIONS OF THE DIRAC EQUATION IN A STRONG COULOMB FIELD AND INSTABILITY OF THE VACUUM AGAINST POSITRON EMISSION

A. Physics of the Electron-Positron Field in a Strong Coulomb Field

The electronic structure of superheavy atoms has been most thoroughly investigated in recent years.[(1-11)] The spectrum found by Greiner and coworkers[(3)] is shown in Fig. 1. First, it is seen that for $Z_c \sim 170$ the lowest bound state solution, the $1s_{1/2}$ state, reaches the top of the negative energy continuum and for further increase of Z ceases to be part of the point spectrum, as will be shown below. At this value of Z, the next higher state, the $2p_{1/2}$ state is several hundred KeV higher, though it, too, descends rapidly with Z and becomes "critical" at $Z \sim 185$. The many-body interpretation of this phenomenon is also well-understood. When the K-shell orbit reaches the energy $-m$ ($\hbar=c=1$), a state of a single such electron is degenerate in energy with a state of two K electrons and a positron at rest (neglecting the electron-electron interaction). If the nuclear charge is increased adiabatically through the critical value $Z_c \sim 170$, but the K shell remains filled throughout this process, then nothing is observed. If there is, however, a hole in the K shell, then beyond $Z = Z_c$, the system will decay into a state of a filled K-shell and a free

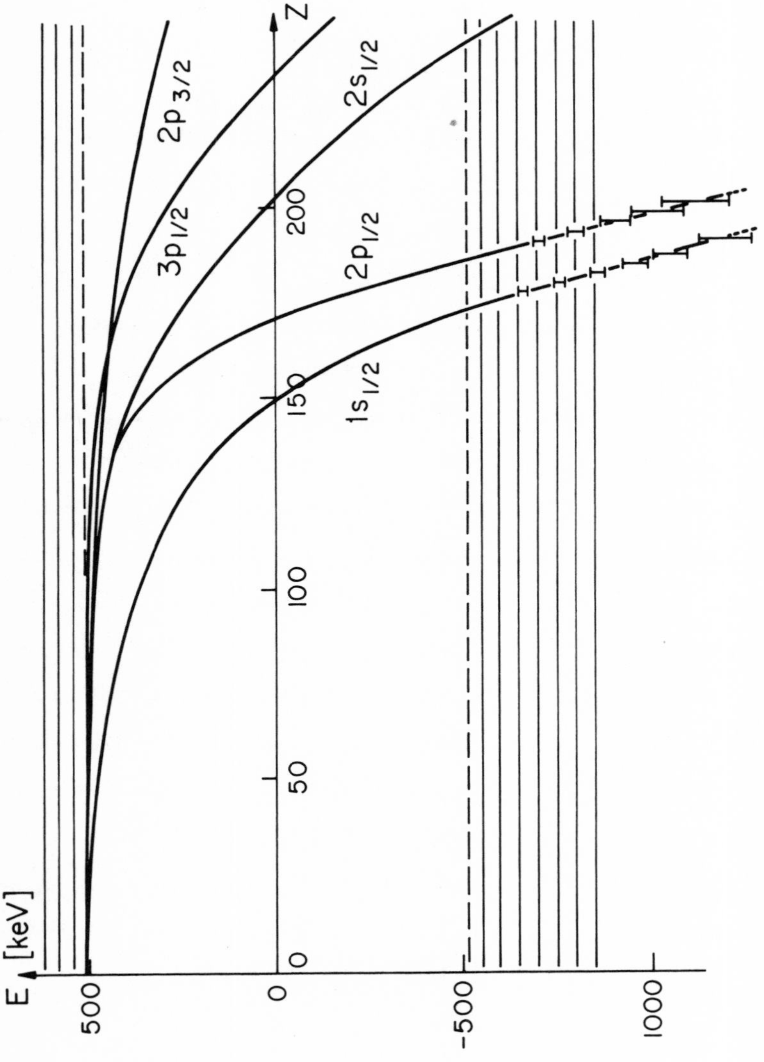

Fig. 1. Energy levels of an electron in the Coulomb field of the extended charge distribution of a nucleus (from ref. 3).

positron. If the K-shell is completely empty, two positrons will materialize at infinity. The only questionable aspect of these remarks concerns the meaning of "filled K-shell" when there is no bound K orbit. This is taken care of by a new definition of the vacuum state.

When bound orbits of negative (total) energy occur, it is sensible to redefine the vacuum state as the state in which all negative energy orbits, bound or unbound, are occupied, since this has a lower energy than the usual vacuum. With this definition, the vacuum is stable for all Z. Moreover, in the new description, a deeply bound hole in the K-orbit is interpreted as a loosely bound positron. When the hole reaches energy -m, the positron reaches energy m and beyond that decays into the continuum. The previous vacuum contains for $Z < Z_c$ two loosely bound positrons which are ultimately available for decay. It is a (crucial) detail of the Dirac equation that as the nuclear charge increases the effective potential binding the positron weakens, ultimately failing to bind. We then encounter standard resonance behavior. On Fig. 1 is marked the width of the resonance that would be observed ideally in the scattering of positrons of suitable energy by the superheavy nucleus.

For the mathematical description of the behavior of the system in the neighborhood of $Z = Z_c$, decisive simplification has been achieved by the introduction of a reduced Hamiltonian. In the remainder of this section we shall give a brief description of this model and its properties.

B. The Reduced Hamiltonian

We are interested in studying the solutions of the Hamiltonian

$$H = \int \psi_\alpha^\dagger(\underline{r})\ [\underline{\alpha}\cdot\underline{p} + \beta m + V(r)]_{\alpha\beta}\ \psi_\beta(\underline{r})\ d^3r$$

$$\equiv \int \psi^\dagger\ H\ \psi\ d^3r, \tag{2.1}$$

$$\{\psi_\alpha(\underline{r}),\ \psi_\beta^\dagger(\underline{r}')\} = \delta_{\alpha\beta}\ \delta(\underline{r}-\underline{r}'), \tag{2.2}$$

where V(r) may be a superstrong (basically Coulomb) potential, also incorporating, possibly, some effects of vacuum polarization (see subsection F). We shall divide V(r), the Coulomb field of the extrapolated charge distribution of the nucleus, into two parts,

$$V(r) = V_o + V_1, \tag{2.3}$$

where the strength of V_o is no more than critical and may be sub-critical; whether the additional V_1 carried us beyond critical is to be specified for each instance.

We divide the solutions of the one-particle

Hamiltonian $H_o = \alpha \cdot \underline{p} + \beta m + V_o(r)$ into three sets $u^{(q)}(\underline{r})$, $u^{(B)}(\underline{r})$, $v^{(q)}(\underline{r})$ designating positive energy continuum states (energies $(m+\varepsilon(q))$), bound states (energies $(-m+\varepsilon_B)$, measured from $-m$!) negative energy continuum states $(-m-\varepsilon(q))$. Following the conventional quantization procedure, we expand in terms of fermion operators a_q, a_B, $b_q^{\dagger}$,

$$\psi_\alpha(r) = \sum_q a_q u_\alpha^{(q)}(\underline{r}) + \sum_B a_B u_\alpha^{(B)}(\underline{r}) + \sum_q b_q^{\dagger} v_\alpha^{(q)}(\underline{r}), \qquad (2.4)$$

where the vacuum is defined by the conditions

$$a_q|vac\rangle = a_B|vac\rangle = b_q|vac\rangle = 0. \qquad (2.5)$$

With the help of (2.4) and of its Hermitian conjugate, and of (2.5), we find, discarding the usual infinite constant,

$$\hat{H} = \sum_q (m+\varepsilon(q))\,(a_q^{\dagger}a_q + b_q^{\dagger}b_q) + \sum_B(-m+\varepsilon_B)\, a_B^{\dagger}a_B + \{\sum_{Bp} U_{Bp}\, a_B^{\dagger}b_p^{\dagger} + \sum_{pp'} U_{pp'}\, a_p^{\dagger}b_{p'}^{\dagger} + \sum_{Bp} V_{Bp}\, a_B^{\dagger}a_{p'} + \text{h.c.}\}$$

$$+ \sum_{BB'} V_{BB'} a_B^{\dagger} a_{B'} + \sum_{pp'} V_{pp'} a_p^{\dagger} a_{p'}$$

$$- \sum_{pp'} W_{pp'} b_{p'}^{\dagger} b_p , \qquad (2.6)$$

where the obvious definitions of the matrix elements U, V, W obtain, e.g.

$$U_{Bp} = (u^{(B)} | V_1 | v^{(q)}). \qquad (2.7)$$

We are interested mainly in the fate of the pair of bound 1s states for which $\varepsilon_B \to 0$ as V_1 increases. If we attempt to apply perturbation theory to the K-orbit, it is natural first to include the diagonal contributions of the last three terms of (2.6) with the unperturbed Hamiltonian. Next, we see that the only unperturbed states nearby in energy to the one-electron state are those with two 1s electrons and one slowly moving positron corresponding to the perturbation U_{Bp}. These latter states may be made as nearly degenerate with the bound state as we wish by taking V_o sufficiently close to critical. The size of V_1 is also at our disposal and can be made as small as we please. Outside of the above class of perturbations, energy denominators are at least several hundred KeV (to the next higher bound state, $2p_{1/2}$, see Fig. 1).

We are thus able to conclude[2] that the behavior of the lowest bound state in the vicinity of critical field strength can be determined by studying a reduced Hamiltonian, H_r, of the form

$$H_r = \sum_{\sigma} \varepsilon\, a_{\sigma}^{\dagger} a_{\sigma} + \sum_{q\sigma} \varepsilon(q)\, b_{q\sigma}^{\dagger} b_{q\sigma}$$

$$\sum_{p\sigma} (U_p a_{\bar{\sigma}}^{\dagger} b_{p\sigma}^{\dagger} + U_p^{*} b_{p\sigma} a_{\bar{\sigma}}), \qquad (2.8)$$

where $\sigma = \pm$ now describes the two possible occupants of the lowest bound states by the projection of their total angular momentum ($\bar{\sigma}=-\sigma$). When necessary, we shall remember that

$$\varepsilon = \varepsilon_o + \varepsilon_1, \qquad (2.9)$$

and that both U_p and ε_1 are proportional to V_1. We have, again, dropped an additive constant, taking into account that for H_r, the difference between the number of bound electrons and the number of positrons is a constant of the motion. Finally, the apparent omission of the last term of (2.6) may be justified by the assumption that the part of H_r depending on continuum positron states has been pre-diagonalized. This involves only a change in the value of U_p.

C. The One-Electron Sector

We now look for eigenstates of H_r of the form

$$|\Psi_\sigma\rangle = A_o a_\sigma^\dagger |vac\rangle + \sum_p A_p a_\sigma^\dagger a_{\bar{\sigma}}^\dagger b_{p\sigma}^\dagger |vac\rangle. \tag{2.10}$$

The attendant Schrodinger equation

$$H_r |\Psi_\sigma\rangle = W |\Psi_\sigma\rangle \tag{2.11}$$

is equivalent to the coupled equations

$$(W-\varepsilon)\, A_o = \sum_p U_p^* A_p, \tag{2.12a}$$

$$[W - 2\varepsilon - \varepsilon(p)]\, A_p = U_p\, A_o. \tag{2.12b}$$

We look first for bound state solutions for which $W_1 - 2\varepsilon - \varepsilon(p) \neq 0$. Solving for A_p from (2.12b) and substituting in (2.12a), we obtain the eigenvalue condition

$$x = F(x-\varepsilon), \tag{2.13a}$$

$$F(y) = \sum_p \frac{|U_p|^2}{y - \varepsilon(p)} \equiv \int_0^\infty dE\, \frac{|V_E|^2}{y - E}, \tag{2.13b}$$

$$x \equiv W_1 - \varepsilon. \tag{2.13c}$$

It is known[4] that for the case of interest $\varepsilon_o = 0$,

$$|V_E|^2 \sim \exp[-2\pi Z\alpha m/p], \tag{2.14}$$

(where Z is the atomic number and α the fine-structure constant), and thus vanishes very rapidly with ε. This is essentially the result of the strong repulsion by the nucleus of very slowly moving positrons. If we were to consider instead a model with short range forces, we can still expect $|V_E|^2$ to vanish like a (fractional) power of E, and therefore in any event the quantity F(0) exists and is negative.

There are two cases to be considered, distinguished in Figs. (2a) and (2b). In (2a), $\varepsilon > 0$. Since $F(-\varepsilon) < 0$, and $F(y) \to 0$ as $y \to -\infty$, (2.13a) has a solution $x = W_1 - \varepsilon < 0$, but as we are about to realize

$$0 < W_1 < \varepsilon. \tag{2.15}$$

Consider the case $\varepsilon < 0$ (and $\varepsilon_o = 0$). In this case $F(x-\varepsilon)$ exists only for $x \leq \varepsilon$. As shown in the figure,

$$|F(0)| < |\varepsilon| \tag{2.16}$$

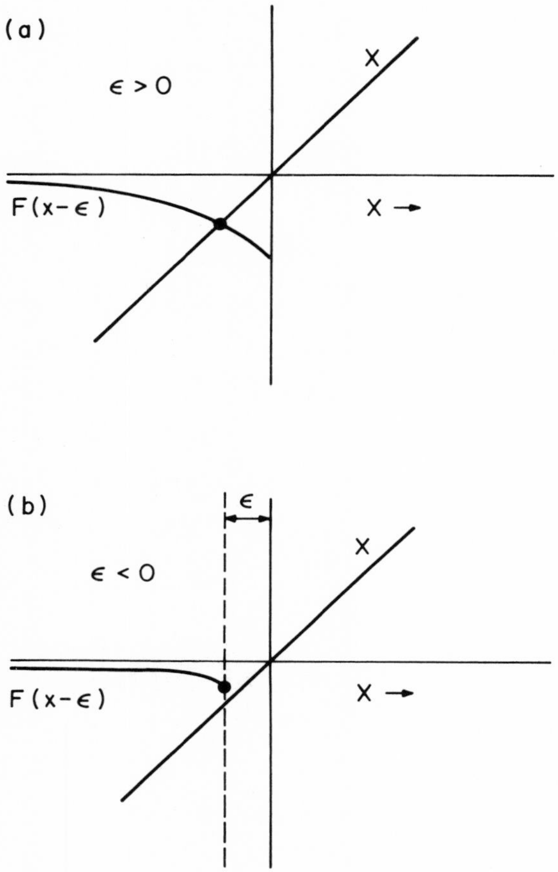

Fig. 2a. Illustration that the equation $F(\chi-\varepsilon) = \chi$ has a real solution for $\varepsilon > 0$, indicating a stable one-electron bound state, $1s_{1/2}$.

Fig. 2b. Illustration that for $\varepsilon < 0$, the same equation has no real solution, indicating an unstable one-electron state $1s_{1/2}$.

and therefore no real root exists. This is obvious for V_1 small enough, since then $|\varepsilon| \sim |V_1|$, whereas $|F(0)| \sim |V_1|^2$. These considerations represent the simplest proof that for $Z > Z_c$, the critical value, a bound 1s state is no longer a member of the point spectrum.

Thus, if we have a bound state, it is represented by the vector

$$|\Psi_\sigma> \rightarrow \begin{pmatrix} A_o \\ \\ A_p \end{pmatrix}, \qquad (2.17a)$$

$$A_p = \frac{U_p A_o}{W - 2\varepsilon - \varepsilon(p)}, \qquad (2.17b)$$

$$|A_o|^{-2} = 1 + \sum_p \frac{|U_p|^2}{(W-2\varepsilon-\varepsilon(p))^2}, \qquad (2.17c)$$

the last being the normalization condition.

There are also scattering solutions of the form (2.10) to be considered. We first remark that for <u>any</u> model considered above, in consequence of the Pauli principle the unperturbed state with two electrons $a_\sigma^\dagger a_{\bar{\sigma}}^\dagger|vac>$ is a bound eigenstate with energy 2ε. The form (2.10) can then represent eigenstates with the boundary condition: positron incident on the state of two electrons, energy $W(q) = \varepsilon(q) + 2\varepsilon$. In this case we write

$$|\Psi_{\sigma q}> \rightarrow \begin{pmatrix} A_{oq}^{(+)} \\ A_{pq}^{(+)} \end{pmatrix} \tag{2.18a}$$

and find for the outgoing wave solutions of the reduced Hamiltonian

$$A_{oq}^{(+)} = \frac{U_q^*}{\varepsilon_+(q) + \varepsilon - F(\varepsilon_+(q))} \tag{2.18b}$$

$$A_{pq}^{(+)} = \delta(p-q) + \frac{U_q^* U_p}{[\varepsilon_+(q) - \varepsilon(p)]} \frac{1}{[\varepsilon_+(q) + \varepsilon - F(\varepsilon_+(q))]} \tag{2.18c}$$

where $\varepsilon_+(q) = \varepsilon(q) + i\eta$, η positive and small. The normalization is such that

$$<\Psi_{\sigma q}|\Psi_{\sigma q'}> = \delta(q-q'), \tag{2.19}$$

as one verifies by straightforward algebra. The scattering solutions are of course orthogonal to the bound state, if the latter exists.

In the event that we are supercritical, the denominator

$$\varepsilon(q) + \varepsilon - F_{+}(\varepsilon(q)) \equiv \varepsilon(q) + \varepsilon - F_{1}(\varepsilon(q))$$

$$+ i\pi \, |V_{\varepsilon(q)}|^2 \, , \qquad (2.20)$$

where F_1 represents the principal value sum, can have a vanishing real part at the resonance energy

$$\varepsilon_R = |\varepsilon| + F_1(\varepsilon_R), \qquad (2.21)$$

because $|F_1(0)| < |\varepsilon|$ and $F_1(0) < 0$; the resonance is, of course, the analytic continuation of the bound 1s state. When there is a true negative energy bound state, there is correspondingly no associated resonance scattering, as can be verified.

We remark finally on the completeness relation satisfied by our solutions. These read (bound state present)

$$|A_o|^2 + \sum_q |A_{oq}^{(+)}|^2 = 1, \qquad (2.22a)$$

$$A_p A_o^* + \sum_q A_{pq}^{(+)} A_{oq}^{(+)*} = 0, \qquad (2.22b)$$

$$A_p A_{p'}^* + \sum_q A_{pq}^{(+)} A_{p'q}^{(+)*} = \delta(p-p'). \qquad (2.22c)$$

If there is no bound state, the first term is missing. The proof of these relations is to be found in ref. 11.

D. Vacuum State Sector

We turn next to the study of eigenstates of the form

$$|\Psi_o> = C_o|vac> + \sum_{p\sigma} C_p\, a_{\overline{\sigma}}^{\dagger}\, b_{p\sigma}^{\dagger}\, |vac>$$

$$+ \sum_{pp'} C_{pp'}\, (a_-^{\dagger} b_{p'+}^{\dagger})\, (a_+^{\dagger} b_{p'}^{\dagger})\, |vac>, \quad (2.23)$$

which includes the vacuum state. We shall quote only the results for this case. We can show that the condition for a stable vacuum state having the same charge as |vac> is the same as the conditions (2.13) for the existence of a stable one-electron state, consideration being given to an appropriate change in the definition of the energy. Thus these two states are stable or unstable together. In the reformulation given below this will be obvious, having the meaning described in subsection A.

E. Definition of a Stable Vacuum and Reformulation of Theory

In the previous two subsections we have

found that in supercritical fields, the state of two electrons bound in the K-shell is stable, but the vacuum and one-electron states are not. A solution to this dilemma is fairly immediate[6,10]. Within the present context the solution requires that we choose the state with a fully occupied K-orbit as the vacuum. Within the context of the original full Hamiltonian, this is a modification of Dirac's original quantization of the electron field, in which the vacuum was the state with the fully occupied sea of negative energy continuum states. In general, we propose that a proper definition of the vacuum includes as well the filling of negative energy bound states. This redefinition simplifies the study of those bound states which encroach upon the top of the negative energy continuum.

The redefinition is effected in practice (as usual) by the interchange of creation and annihilation operators. We define

$$b_\sigma^\dagger = a_{\overline{\sigma}},$$

$$b_\sigma = a_{\overline{\sigma}}^\dagger. \qquad (2.24)$$

We say that $b_\sigma^\dagger$ creates a bound positron, i.e., the absence of a deeply bound electron is reinterpreted as the presence of a weakly bound positron. This is seen by rewriting the reduced Hamiltonian (2.8), which to an additive constant becomes

$$H_r' = -\varepsilon \sum_\sigma b_\sigma^\dagger b_\sigma + \sum_p \varepsilon_p b_{p\sigma}^\dagger b_{p\sigma}$$

$$- \sum_{p\sigma} U_p b_{p\sigma}^\dagger b_\sigma - \sum_{p\sigma} U_p^* b_\sigma^\dagger b_{p\sigma}. \qquad (2.25)$$

We now have a theory exclusively of positrons. If $\varepsilon > 0$, there can be bound positrons, which, as we have seen, remain bound under perturbation. If $\varepsilon < 0$, the unperturbed system has a bound state in the continuum which becomes a resonance under the perturbation. Thus, if we use the new vacuum and the language of positrons, the situation concerning the behavior of the K-shell in supercritical fields loses completely any aura of mystery and is reduced to the prototype problem of a weakly bound state transforming under further change in potential into a resonance, and the model is equivalent to one considered long ago by Fano.[24]

It is unnecessary to calculate the eigenstates of (2.25). They may be read off from the one-electron solutions of subsection C, changing only a few signs, in accordance with the change of sign in the interaction terms of (2.25). The solutions will be written in terms of the operators which diagonalize the Hamiltonian. Including the case of a bound system we introduce operators β_σ, $\beta_{q\sigma}$, given by the linear relations

$$\beta_\sigma = A_o\, b_\sigma - \sum_p A_p\, b_{p\sigma}, \qquad (2.26a)$$

$$\beta_{q\sigma} = -\, A_{oq}^{(+)}\, b_\sigma + \sum_p A_{pq}^{(+)}\, b_{p\sigma}, \qquad (2.26b)$$

where A_o and A_p are given by (2.17) and $A_{oq}^{(+)}$, $A_{pq}^{(+)}$ are given by (2.18b) and (2.18c). Instead of W_1 we shall use the eigenvalue

$$\omega = W_1 - 2\varepsilon. \qquad (2.27)$$

The fermion operators (2.26) are the solution of the equations of motion and render the Hamiltonian diagonal, e.g.,

$$H_r' = \sum_\sigma \omega\, \beta_\sigma^\dagger \beta_\sigma + \sum_{q\sigma} \varepsilon(q)\, \beta_{q\sigma}^\dagger \beta_{q\sigma}. \qquad (2.28)$$

The orthogonality and completeness relations are unaffected by the phase changes indicated in (2.26). The completeness relations allow us to invert the relations (2.26),

$$b_\sigma = A_o^*\beta_\sigma - \sum_q A_{oq}^{(+)*}\beta_{q\sigma}, \qquad (2.29a)$$

$$b_{p\sigma} = -A_p^*\beta_\sigma + \sum_q A_{pq}^{(+)*}\beta_{q\sigma}. \qquad (2.29b)$$

These transformations together with the orthogonality and completeness relations permit the discussion of any question of physical interest. For example we may follow a standard treatment[25] to derive the approximate exponential decay of the probability $|\langle vac|b_\sigma \exp[-H'_r t]\, b_\sigma^\dagger|vac\rangle|^2$ and it follows as well from (2.29a) that $|A_{oq}^{(+)}|^2$ describes the energy spectrum of the decay positrons (when there is no bound state).

F. Effect of Vacuum Polarization

We believe that the additional effects of vacuum polarization should have only a minor influence on the theory described up to this point. Consider the new definition of the vacuum. Divide the solutions of the Dirac equation into the positive energy ones $u_\alpha^{(\nu)}(r)$ and the negative energy ones $v_\alpha^{(\lambda)}(r)$, including bound states. The potential $V(r)$ in Eq. (2.1) and sequel should then be calculated self-consistently from the equation ($V=eA_o, \psi=u$ or v)

$$E_\nu\, \psi^{(\nu)}(\underline{r}) = [\underline{\alpha} \cdot \underline{p} + \beta m + eA_o(\underline{r})]\, \psi^{(\nu)}(\underline{r}), \tag{2.30a}$$

$$\nabla^2 A_o(\underline{r}) = -\,\rho_{ext}(\underline{r}) - \rho_{vac}(\underline{r}), \tag{2.30b}$$

where ρ_{ext} is the charge density of the "external

field" (see below) and $\rho_{vac}(r)$ is the vacuum polarization charge.

$$\rho_{vac}(r) = \sum_{\alpha} \langle vac|\ |e|\ \tfrac{1}{2}\ [\psi_\alpha(r),\ \psi_\alpha^\dagger(r)]|vac\rangle$$

$$= \tfrac{1}{2}\ |e|\ \{\sum_{\alpha\nu} |u_\alpha^{(\nu)}(r)|^2$$

$$- \sum_{\alpha,\lambda} |v_\alpha^{(\lambda)}(r)|^2\}. \qquad (2.31)$$

By ρ_{ext}, we mean a quantity which includes the contribution of the negative energy bound electrons. This implies that in a linearized approximation below the critical field strength, the right hand side of (5.1b) is almost independent of the definition of the vacuum, since

$$\rho_{ext} = \rho_{ext}^{(0)} - \sum_{B,\alpha} |e|\ |v_\alpha^{B}(r)|^2, \qquad (2.32)$$

where $\rho_{ext}^{(0)}$ is the conventional definition of the external charge. Above the critical value of Z, this separation is no longer strictly possible for the lowest bound states, the corresponding terms disappearing from (2.31).

We note that with the new definition of the vacuum, ρ_{vac} has the same structure as in the weak coupling case (positive energies with one sign, negative energies with another). This suggests that the technique of Wichmann and Kroll[26] may continue to be applicable to the evaluation of vacuum polarization. In consequence, the non-linear problem (5.1) may prove tractable, though not trivial. There is no mechanism here to alter the essentials of the previous discussion, only the details. We believe that the problem posed by (2.30) is worthy of investigation.

III. KLEIN-GORDON EQUATION IN A STRONG COULOMB FIELD. INSTABILITY AGAINST CHARGED PAIR CONDENSATION

A. Forms of the K.G. Equation

We study (with $\hbar=c=1$) the equation

$$(E-V)^2 \psi(r) = (p^2+m^2) \psi(r), \tag{3.1}$$

where $\psi(r)$ is a complex scalar function and $V(r)$ is the potential energy of a negatively charged scalar particle of mass m (henceforth called pion) in the field of a fixed extended charge distribution. For example, we may take

$$V(r) = - \frac{Ze^2}{r} f(r), \tag{3.2}$$

$$f(r) = 1 - e^{-\kappa r} \tag{3.3}$$

where $\kappa^{-1} \sim R$, the nuclear radius.

The formal properties of the equation are more easily studied if we introduce a formalism first order in the energy (time derivative)[27]. In terms of the two-component vector

$$\Phi = \begin{pmatrix} \psi \\ (E-V)\psi \end{pmatrix}. \tag{3.4}$$

Eq. (3.1) becomes

$$E\, \tau_1\, \underline{\Phi} = H\underline{\Phi}, \tag{3.5}$$

$$H = \frac{1}{2}(1+\tau_3)(p^2+m^2) + \frac{1}{2}(1-\tau_3) + \tau_1 V. \tag{3.6}$$

Here τ_i, i = 1,2,3 are the usual Pauli spin matrices. Since H is Hermitian, we recognize that the fundamental scalar product is the integral of the density

$$d_{ab}(r) = \underline{\Phi}_a^{\dagger}(r)\, \tau_1\, \underline{\Phi}_b(r), \tag{3.7}$$

providing an orthogonality theorem for two solutions of (3.5) belonging to different energies.

For the norm of any given solution, we compute

$$\int \underline{\Phi}^{\dagger} \tau_1 \underline{\Phi} = 2 \int \psi^{*}(E-V)\psi. \tag{3.8}$$

Remembering the sign of V, Eq. (3.2), we see that (3.8) is certainly positive for $E \geq 0$. On the other hand, for negative energy continuum solutions, the norm must be negative, by analytic continuation from the limit $V = 0$. Thus if we choose the original K.G. scalar function to be normalized according to

$$\int |\psi|^2 = [2|E - \langle V\rangle|]^{-1}, \tag{3.9}$$

where $\langle V\rangle = \int \psi^{*} V\psi / \int \psi^{*}\psi$, we see that, in general, the solution set for given V divides into two subsets $\underline{\Phi}_p$ and $\underline{\Phi}_n$ characterized as follows:

$$\int \underline{\Phi}_p^{\dagger} \tau_1 \underline{\Phi}_p = 1, \; E_p - \langle V\rangle_p > 0,$$

$$\int \underline{\Phi}_n^{\dagger} \tau_1 \underline{\Phi}_n = -1, \; E_n - \langle V\rangle_n < 0. \tag{3.10}$$

The considerations above obviously fail, should there occur an eigenvalue for which the inequalities (3.10) are replaced by an equality. The resolution of this difficulty, carried out below, represents the basic goal of this part of our work.

For the normally occurring situations for which (3.10) applies, the completeness relation for the solutions of (3.5) takes the dyadic form

$$\sum_p \underline{\Phi}_p(r)\,[\underline{\phi}_p{}^{\dagger}(r')\,\tau_1] - \sum_n \underline{\Phi}_n(r)\,[\underline{\Phi}_n{}^{\dagger}(r')\,\tau_1]$$

$$= I\delta(r-r'), \qquad (3.11)$$

and I is the unit two by two matrix.

A third form of the K.G. equation is sometimes useful because of its analogy with the non-relativistic Schrodinger equation. For this form, we write

$$\varepsilon_{eff}\psi = (p^2/2m)\psi + V_{eff}\psi, \qquad (3.12)$$

where

$$\varepsilon_{eff} = E'[1 + (E'/2m)],$$

$$V_{eff} = V[1 + (E'/m) - (V/2m)],$$

$$E = E' + m. \qquad (3.13)$$

B. Bound State Spectrum and Approach to the Critical Point[12]

For full understanding of the situation when Z becomes very large, we consider together the solutions for both negative and positive pions ($\pi_\pm$). The situation is illustrated schematically in Fig. 3. For a potential energy of the form (3.2), the curve marked E_- represents in its solid part the lowest bound state of a π_-. Two special values of Z are to be noted. For $Z > Z_o$, there emerges from the negative energy continuum a new bound state branch for π_- which meets the branch E_- at a point of vertical tangency, $Z = Z_c$. For $Z > Z_c$ there is simply no bound state solution corresponding to this branch. The curve marked E_+ is a reflection of E_- with respect to the abscissa $E = 0$ and represents a solution branch for π_+.

We shall now indicate the derivation of these results from the K.G. equation. To understand their physical implication we must have resort to the quantum field theory described in the next section.

It follows most easily from (3.1) that

$$\psi(r;E,e) = \psi(r;-E,-e). \tag{3.14}$$

This shows that if branch E_- occurs for π_-, branch E_+ occurs for π_+. Next consider the equation for π_+ near zero binding energy. From (3.13), remembering to change the sign of the charge, we have

$$V_{eff}(\pi_+, E'\sim 0) \cong |V(r)| - [V^2/2m]. \tag{3.15}$$

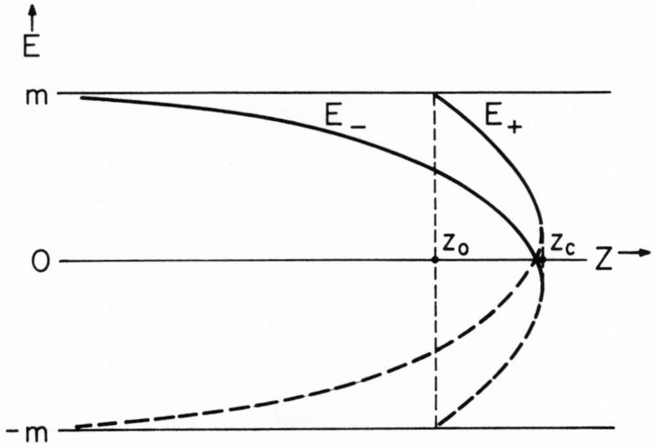

Fig. 3. Schematic representation of the lowest bound state branch for a negative pion in the Coulomb field of a superheavy nucleus (neglecting strong interactions). The branch for the positive pion related by charge conjugation is also shown.

For Z large enough, the second term must dominate and give ($Z>Z_o$) a π_+ bound state -- even in the field of a positively charged nucleus, -- as shown in the solid part of the branch E_+. The combination of (3.14) and (3.15) establishes the qualitative character of Fig. 3.

It is furthermore easy to see from the considerations of subsection A that points on the solid parts of curves $E_\pm$ correspond to solutions with positive norm whereas dashed portions represent solutions with negative norm. The meeting point must therefore be one at which $E = \langle V\rangle$. Now consider Eq. (3.1) for $V \to \lambda V$ and form the expectation value,

$$\langle (E-\lambda V)^2\rangle = \langle p^2\rangle + m^2. \qquad (3.16)$$

With the help of the K.G. equation, the first derivative of (3.16) at $\lambda = 1$ becomes

$$[E - \langle V\rangle]\, dE/d\lambda = \langle EV\rangle - \langle V^2\rangle. \qquad (3.17)$$

At the point $E = \langle V\rangle$, the right hand side cannot vanish. Therefore we must have $dE/d\lambda \to \infty$, a point of vertical tangency.

Let $E_- = -\mu$ represent this point occurring at $Z = Z_c$. Then we see that for the branch E_+, we have $E_+ = \mu$. Therefore in a quantum field theory we could produce an indefinite number of π_+ pairs without energy cost (in the absence of other interactions). This apparent instability will be dealt with in the next section.

Let us estimate the value of Z_c predicted by our model, using the pion mass. This can be done with the help of a virial theorem. From the statement

$$<[\underline{r} \cdot \underline{p}, \{(E-V)^2 - p^2 - m^2\}]> = 0, \qquad (3.18)$$

we derive

$$<p^2> = <(r.\nabla V)\ (E-V)>. \qquad (3.19)$$

Combining this with Eqs. (3.16) (λ=1) and (3.2,3) we derive ($\alpha = e^2/\hbar c$)

$$E[E-<V>] = m^2 - Z\alpha E<f'(r)> + Z\alpha<f'(r)V>. \qquad (3.20)$$

The correct order of magnitude for Z_c should be obtained if we examine (3.20) for E = 0. Thus we have

$$m^2 = (Z\alpha)^2\kappa\ <e^{-\kappa r}/r> \sim \frac{(Z\alpha)^2}{R^2}, \qquad (3.21)$$

using the essential fact that the pion wave-function is largely confined to the nuclear interior. If we put $R^2 = (10^2/m^2)$ we find $Z\alpha \sim 10$, which is roughly consistent, but unfortunately renders our problem somewhat academic, at least under present conditions. (If we had a pion of electronic mass, Z_c would be reduced by an order of magnitude.) Thus, if there is to be an observable pion condensation phenomena, we shall have to look for its origin in the nuclear forces. This problem is considered in secs. V and VI.

Returning now to the technical aspects of our problem, we notice that at $Z = Z_c$, there is only one solution remaining of the two we had for $Z < Z_c$. To understand the approach to the limit, we define in terms of the two solutions of interest $\underline{\Phi}_{\pm}$ ($\pm$ referring to norm), two new uncharged linear combinations

$$\underline{\Phi}_e = (1/\sqrt{2})\ (\underline{\Phi}_+ + \underline{\Phi}_-),$$

$$\underline{\Phi}_o = (1/\sqrt{2})\ (\underline{\Phi}_+ - \underline{\Phi}_-). \tag{3.22}$$

With

$$d_e(r) = \underline{\Phi}_e^{\dagger}(r)\ \tau_1 \underline{\Phi}_e(r), \tag{3.23}$$

etc., we have straightforwardly

$$\int d_e = \int d_o = 0, \tag{3.24}$$

but

$$\int d_{eo} = \int d_{oe} = 1. \qquad (3.25)$$

It may be instructive to exhibit expressions for these densities. In the following we make use of the orthogonality between $\underline{\Phi}_+$ and $\underline{\Phi}_-$, which can be expressed in the form, assuming real wave functions,

$$E_+ + E_- = 2V_{+\,-}, \qquad (3.26)$$

where

$$V_{ab} = (\psi_a, V\psi_b)/[(\psi_a,\psi_a)(\psi_b,\psi_b)]^{1/2} \qquad (3.27)$$

and (as will be needed below) $V_{aa} = V_a$. We also record our expressions using unit normalization for ψ_a. We then have

$$d_e(r) = \frac{1}{2}\left\{\psi_+^2 - \psi_-^2 + \frac{(V_+-V(r))\psi_+^2}{|E_+ - V_+|} + \frac{(V_--V)\psi_-^2}{|E_- - V_-|} + \frac{2(V_{+-}-V)\psi_+\psi_-}{[|E_+ - V_+|\;|E_- - V_-|]^{1/2}}\right\}, \qquad (3.28)$$

$$d_{oe}(r) = \frac{1}{2}\left\{\psi_+^{\,2} + \psi_-^{\,2} + \frac{(V_- - V)\psi_-^{\,2}}{|E_- - V_-|} - \frac{(V_+ - V)\psi_+^{\,2}}{|E_+ - V_+|}\right\}, \tag{3.29}$$

$$d_{o}(r) = \frac{1}{2}\left\{\psi_+^{\,2} - \psi_-^{\,2} + \frac{(V_+ - V)\psi_+^{\,2}}{|E_+ - V_+|} - \frac{(V_- - V)\psi_-^{\,2}}{|E_- - V_-|}\right\}. \tag{3.30}$$

From these expressions, we verify (3.24) and (3.25). From (3.28) we see that the limit $Z \to Z_c$ cannot be taken, since, if we define

$$\alpha = \frac{1}{2}\,\{|E_+ - V_+| + |E_- - V_-|\}, \tag{3.31}$$

d_e behaves like $O(\alpha^{-1})$. By contrast, d_{oe} and d_o remain finite as $\alpha \to 0$.

IV. QUANTIZATION OF THE KLEIN-GORDON EQUATION IN A STRONG COULOMB FIELD. STABILIZATION OF THE VACUUM BY NON-LINEAR INTERACTIONS

A. Quantization Below the Critical Point(28,29)

This is achieved most elegantly in an external field by use of the first-order formalism. We adopt the Lagrangian

$$L(t) = \int d^3r \, [i\underline{\Phi}^{\dagger}_{op}(r,t)\tau_1 \, \dot{\underline{\Phi}}_{op}(r,t) - \underline{\Phi}^{\dagger}_{op} \, H\underline{\Phi}_{op}]. \tag{4.1}$$

This yields the canonical momentum

$$\underline{\pi}_{op}(r,t) = i\underline{\Phi}^{\dagger}_{op}(r,t)\tau_1. \tag{4.2}$$

Thus the required commutation relations are

$$[\underline{\Phi}_{op}(r,t), \, \underline{\Phi}^{\dagger}_{op}(r',t)] = \tau_1 \, \delta(r-r'). \tag{4.3}$$

The Hamiltonian which follows from (4.1) and (4.2) is

$$H = \int \underline{\Phi}^{\dagger}_{op} \, H\underline{\Phi}_{op}. \tag{4.4}$$

To this we adjoin our candidate for total charge operator, namely

$$Q = -|e| \int \underline{\Phi}^{\dagger}_{op} \, \tau_1 \, \underline{\Phi}_{op}. \tag{4.5}$$

It is understood that H and Q are to be taken in normal form with respect to the vacuum state to be defined below.

The completeness relation (3.11) will guarantee us a satisfactory quantum theory when used in conjunction with the expansion

$$\underline{\Phi}_{op}(r) = \sum_p a_p \, \underline{\Phi}_p(r) + \sum_n b_n{}^\dagger \underline{\Phi}_n(r), \qquad (4.6)$$

if we assume that

$$[a_p, a_{p'}{}^\dagger] = \delta_{pp'} ,$$

$$[b_n, b_{n'}{}^\dagger] = \delta_{nn'} , \qquad (4.7)$$

(and, of course, if the a's and b's commute, etc.). Thus (4.6) and (4.7) satisfy (4.3). Furthermore if the vacuum is annihilated by a_p and b_n, we find for (4.4) and (4.5),

$$H = \sum_p E_p \, a_p{}^\dagger a_p + \sum_n |E_n| \, b_n{}^\dagger b_n, \qquad (4.8)$$

$$Q = - |e| \, \{\sum_p a_p{}^\dagger a_p - \sum_n b_n{}^\dagger b_n\}. \qquad (4.9)$$

As long as $Z < Z_c$, this represents an unequivocally satisfactory theory of non-interacting bosons of either charge in the external field of a positively charged nucleus. As we approach the point $Z = Z_c$, we arrive, now on a proper quantum basis at the situation already described: the vacuum state is no longer unique, but approaches degeneracy with an infinite set of other states containing various numbers of $\pi_\pm$ of energy $\pm\,\mu$ each.

To remedy this situation, it is sufficient to include in the quantum theory the mutual Coulomb interaction of any produced pions. This will be demonstrated with sufficient rigor below, but the result can be anticipated on physical grounds. Thus as we approach $Z = Z_c$, including the Coulomb interaction of the pions, multi-pion states of small excitation energy can now mix with the previously defined vacuum. But the expectation value of the self-Coulomb interaction of any assembly of charges is positive and since it is quartic in the amplitudes, it must ultimately dominate and prevent collapse. Moreover, if we consider the equation of motion for excitations of negative charge, which is the appropriate generalization of the unquantized K.G. equation, we must find that the charge distribution of the other pions screens the nuclear field on the average. The net result is an effective nuclear field which remains subcritical. A vacuum state is thus stabilized but in terms of the eigenstates of the Hilbert space defined by the expansion (4.6), its description will involve many components.

B. Quantization Above the Critical Point. The Reduced Hamiltonian

Formally the quantization can be carried out precisely as above. We need only replace (4.4) by the Hamiltonian

$$H = \int \underline{\Phi}^{\dagger}_{op} \; \mathcal{H}\underline{\Phi}_{op}$$

$$+ \frac{1}{2} e^2 \int \frac{[\underline{\Phi}^{\dagger}_{op}(r)\tau_1\underline{\Phi}_{op}(r)][\underline{\Phi}^{\dagger}_{op}(r')\tau_1\underline{\Phi}_{op}(r')]}{|r - r'|}. \tag{4.10}$$

Since V in $\mathcal{H}$ is now, by assumption, beyond critical, $|V| > |V_c|$ we choose V_o such that $|V_o| < |V_c|$ and $[|V_c - V_o|/|V_c|] \ll 1$ and write

$$\mathcal{H} = \mathcal{H}_o - \tau_1\Delta,$$

$$V - V_o = -\Delta, \; \Delta > 0. \tag{4.11}$$

We also define

$$\varepsilon_{\pm} \; \tau_1 \; \underline{\Phi}_{\pm} = \mathcal{H}_o \; \underline{\Phi}_{\pm}, \tag{4.12}$$

$$\delta = \frac{1}{2} (\varepsilon_+ - \varepsilon_-) > 0, \tag{4.13}$$

$$\mu_o = -\frac{1}{2} (\varepsilon_- + \varepsilon_+) > 0, \tag{4.14}$$

where these all refer to the near critical eigenvalues.

To establish the stability of the Hamiltonian (4.10), we write

$$\underline{\Phi}_{op}(r) = a\underline{\Phi}_{+}(r) + b^{\dagger}\underline{\Phi}_{-}(r) + \chi_{op}(r). \quad (4.15)$$

For the remainder of the present discussion we drop the terms arising from χ_{op}, since these make reference to the non-dangerous levels only: their absence can effect quantitative details only. We also rewrite the first terms of (4.15) by means of the operators

$$q = (a+b^{\dagger})/\sqrt{2} \ ,$$

$$p = i(a^{\dagger}-b)/\sqrt{2} \ , \quad (4.16)$$

yielding

$$\underline{\Phi}_{op}(r) = q\underline{\Phi}_{e}(r) + ip^{\dagger}\underline{\Phi}_{o}(r), \quad (4.17)$$

where $\underline{\Phi}_e$ and $\underline{\Phi}_o$ are the combinations defined in (3.22). With the aid of definitions (4.11) - (4.14) and the further definitions

$$\Delta_{ab} = \int \underline{\Phi}_a{}^{\dagger}\tau_1 \ \Delta\underline{\Phi}_b \ , \quad (4.18)$$

$$U_{ab,cd} = e^2 \int (\Phi_a{}^{\dagger}\tau_1\Phi_b)\ (\Phi_c{}^{\dagger}\tau_1\Phi_d)'/|r - r'|, \quad (4.19)$$

$$L = b^\dagger b - a^\dagger a = :i(q^\dagger p^\dagger - pq):, \qquad (4.20)$$

we obtain by straightforward transcription (where $U_{eeee} \equiv U_e$, etc.)

$$H_{red} = (\delta - \Delta_{oo})p^\dagger p + (\delta - \Delta_{ee})q^\dagger q + (\mu_o + \Delta_{oe})L$$

$$+ \frac{1}{2} U_e (q^\dagger q)^2 + \frac{1}{2} U_{e,o}\{q^\dagger q, p^\dagger p\}$$

$$+ \frac{1}{2} U_o (p^\dagger p)^2 - \frac{1}{2} \{q^\dagger q, L\} U_{e,oe}$$

$$- \frac{1}{2} \{p^\dagger p, L\} U_{o,oe} + \frac{1}{2} L^2 U_{oe,oe}. \qquad (4.21)$$

The subscript on H reminds us that we are describing only two degrees of freedom.

For $Z - Z_c \sim 1$, the basis defined by (4.11) and (4.12) should be very satisfactory for the diagonalization of the Hamiltonian (4.21). The unperturbed vacuum (we consider the subspace $L = 0$) will mix with states with at most a few pion pairs to yield a new stable vacuum. This is guaranteed by the positive definite character of the quartic terms of H.

For $Z - Z_c \equiv Z' >> 1$, the mixing becomes large and the treatment by means of the basis defined above cumbersome. In this case the stability of the ground state is seen in a classical approximation for which

$$<(q^\dagger q)^2> \cong <q^\dagger q> <q^\dagger q> = (q^2)^2 \qquad (4.22a)$$

$$<q^\dagger q \; p^\dagger p> = <q^\dagger q> <p^\dagger p> = q^2 p^2, \qquad (4.22b)$$

etc.

With

$$\delta - \Delta_{oo} = - B, \; \delta - \Delta_{ee} = - A, \qquad (4.23)$$

we have (L=0)

$$W(q^2,p^2) = <H_{red}> = - Aq^2 - Bp^2$$

$$+ \frac{1}{2} U_e (q^2)^2 + U_{eo} q^2 p^2$$

$$+ \frac{1}{2} U_o (p^2)^2. \qquad (4.24)$$

As will be seen below, we can, without loss of generality, choose $p^2 = 0$. The variation of W with respect to q^2 yields the minimum at

$$q^2 = A/U_e, \qquad (4.25)$$

and

$$W = - \frac{1}{2} A^2/U_e, \qquad (4.26)$$

which is proportional to Z'^2.

Further progress and substantiation of the above simplification depends on recognizing that the variational expression (4.24) may be derived as the expectation value of H_{red} with respect to a coherent trial function

$$|a',b',\theta> = \exp(a^\dagger a'+b^\dagger b' e^{-i\theta})|vac>. \quad (4.27)$$

(This is true insofar as $|a'|^2 >> 1$, $|b'|^2 >> 1$.) With the identifications

$$q^2 = |q_o|^2, \; p^2 = |p_o|^2, \quad (4.28)$$

we have

$$\sqrt{2}\, q_o = a' + b' e^{i\theta}, \quad (4.29)$$

$$\sqrt{2}\, p_o = i(a'-b'e^{-i\theta}). \quad (4.30$$

Average charge neutrality is assured by choosing $a' = b'$. The choice $\theta = 0$ for the arbitrary phase furthermore guarantees $p_o = 0$.

The observations just made suggest that the previous classical approximation can be improved by also varying the operators $a^\dagger$ and $b^\dagger$. The expectation value of H_{red} with respect to a general trial state (4.28) again has the form (4.24) with the altered definitions

$$A = -\int \underline{\Phi}_e^{\dagger} H \underline{\Phi}_e, \qquad (4.31a)$$

$$B = -\int \underline{\Phi}_o^{\dagger} H \underline{\Phi}_o. \qquad (4.31b)$$

To obtain equations to define Φ_e and Φ_o, we vary with respect to $q^2 \underline{\Phi}_e^{\dagger}$ and $p^2 \underline{\Phi}_o^{\dagger}$. These variations must be constrained by the condition of charge neutrality

$$- \mu \langle Q \rangle = - \mu [q^2 \int \underline{\Phi}_e^{\dagger} \tau_1 \underline{\Phi}_e + p^2 \int \underline{\Phi}_o^{\dagger} \tau_1 \underline{\Phi}_o]$$

$$= 0, \qquad (4.32)$$

which, however, still doesn't define the scale of $\underline{\Phi}_e$ and $\underline{\Phi}_o$. We therefore subtract

$$- \delta \, [q^2 \int \underline{\Phi}_e^{\dagger} \tau_1 \underline{\Phi}_o + p^2 \int \underline{\Phi}_o^{\dagger} \tau_1 \underline{\Phi}_e + c.c.]. \qquad (4.33)$$

Thereby we have introduced the Lagrange multipliers μ and δ (cf. (4.12) and (4.14) where μ corresponds to $-\mu_o$). In fact, now carrying out the proposed variations, we obtain the equations

$$H_{eff} \underline{\Phi}_e = \mu \tau_1 \underline{\Phi}_e + \delta \tau_1 \underline{\Phi}_o \qquad (4.34a)$$

$$H_{eff}\, \underline{\Phi}_o = \mu\tau_1\, \underline{\Phi}_o + \delta\tau_1\, \underline{\Phi}_e, \qquad (4.34b)$$

of the form desired, with

$$H_{eff} = H + V_{eff}, \qquad (4.35)$$

$$V_{eff}(r) = e\tau_1 \int |r-r'|^{-1}\, [q^2\, \underline{\Phi}_e{}^{\dagger}(r')\tau_1\, \underline{\Phi}_e(r')$$

$$+ p^2\, \Phi_o{}^{\dagger}(r')\, \tau_1\, \underline{\Phi}_o(r')]$$

$$\equiv e\tau_1 \int |r - r'|^{-1}\, \rho_{eff}(r'). \qquad (4.36)$$

Here the self-consistent Hamiltonian H_{eff} is to be subcritical, but to correspond to a situation in which the levels defined by $\underline{\Phi}_e$ and $\underline{\Phi}_o$ dominate the situation. We want to be in the neighborhood $\delta \sim 1$ (where unity means the energy change upon changing Z by one).

From Eq. (4.34) we can derive the following equations for q^2 and p^2,

$$q^2\, U_e + p^2\, U_{e,o} = A + \delta, \qquad (4.37a)$$

$$q^2\, U_{e,o} + p^2\, U_o = B + \delta. \qquad (4.37b)$$

These equations can all be simplified by the

choice p = 0, but this simplification will not be carried out here.

The procedure just carried out provides the basis for a full discussion of H_{red} in the strong coupling regime $Z' \gg 1$. We shall allude only very briefly to this matter, since numerical studies are in progress and we intend to publish a full account elsewhere. One studies the Hamiltonian

$$H'_{red} = H_{red} - \mu Q, \qquad (4.38)$$

writing the operator $\underline{\Phi}(r)$ as the sum of a classical and a quantum part

$$\underline{\Phi}(r) = \underline{\Phi}_{eff}(r) + \underline{\chi}(r) \ , \qquad (4.39)$$

where

$$\underline{\Phi}_{eff}(r) = q_o \ \underline{\Phi}_e(r) + i \ p_o^{\dagger} \underline{\Phi}_o(r), \qquad (4.40)$$

$$\underline{\chi}(r) = q \ \underline{\Phi}_e(r) + i \ p^{\dagger} \underline{\Phi}_o(r). \qquad (4.41)$$

There follows straightforward but lengthy algebra. To terms of relative order $(Z')^{-1}$, we find

$$H'_{red} \cong -\frac{1}{2} e^2 \int \rho_{eff}(r) \, |r - r'|^{-1} \, \rho_{eff}(r')$$

$$+ \frac{1}{2} e^2 \int \eta_{op}(r) \, |r - r'|^{-1} \, \eta_{op}(r'), \tag{4.42}$$

where

$$\eta_{op}(r) = \Phi_{eff}{}^{\dagger}(r) \, \tau_1 \, \chi(r) + \text{h.c.} \tag{4.43}$$

Since η_{op} is Hermitian, the second term of (4.42) is positive definite. It is also quadratic in the variables q, $q^{\dagger}$, p, $p^{\dagger}$. By a linear canonical transformation it can be brought to the form

$$\alpha P^{\dagger} P + \beta Q^{\dagger} Q, \tag{4.44}$$

where α and β are positive. All terms omitted from (4.42) can be treated by perturbation theory. This treatment then exhibits once more the existence of stable solutions.

V. METHODS FOR THE STUDY OF PION CONDENSATION IN INFINITE MATTER

A. Neutral Pion Condensate

Because it is formally simplest, we shall

first consider the theory of a neutral pion condensate. It is convenient to use the quantum theory associated with the conventional second order wave-equation.

The action integral

$$I = \frac{1}{2}\int d^4x \, [- (\partial_\mu\Phi)^2 - m_\pi{}^2\Phi^2] + \int d^4x \, L^{(int)} \, (\Phi, \partial_\mu\Phi) + I_N, \tag{5.1}$$

where I_N is the action depending on variables other than Φ, leads to the formal equations of motion

$$\partial_\mu^2\Phi - m_\pi^2\Phi = j, \tag{5.2}$$

$$j(x) = -\left(\frac{\partial L^{int}}{\partial\Phi(x)} - \partial_\mu \frac{\partial L^{(int)}}{\partial\partial_\mu\Phi(x)} \right) . \tag{5.3}$$

With the definition

$$\pi(x) = \frac{\partial L}{\partial\dot{\Phi}(x)} = \dot{\Phi}(x) + \frac{\partial L^{int}}{\partial\dot{\Phi}(x)} , \tag{5.4}$$

we compute the Hamiltonian

$$H = \frac{1}{2}\int d^3x\,[\pi^2 + (\nabla\Phi)^2 + m_\pi{}^2\Phi^2]$$

$$- \int d^3x \left[L^{(\text{int})} + \frac{1}{2}\left(\frac{\partial L^{(\text{int})}}{\partial\dot{\Phi}}\right)^2\right]$$

$$+ H_N. \tag{5.5}$$

If, as we assume, the quantization is consistent, we derive the equations of motion

$$\dot{\pi} = (\nabla^2 - m_\pi{}^2)\Phi + \left(\frac{\partial L^{(\text{int})}}{\partial\Phi} - \nabla\cdot\frac{\partial L^{(\text{int})}}{\partial\nabla\Phi}\right),$$

$$\dot{\Phi} = \pi - \frac{\partial L^{(\text{int})}}{\partial\dot{\Phi}}, \tag{5.6}$$

which upon elimination of π return Eq. (5.2). Formally we have

$$\dot{\pi} = -\frac{\delta H}{\delta\Phi},$$

$$\dot{\Phi} = \frac{\delta H}{\delta\pi}. \tag{5.7}$$

The occurrence of a condensate is signaled by a non-vanishing expectation value of Φ and possibly of π. Because $\delta\langle H\rangle = \langle\delta H\rangle$, and considering a c-number variation we conclude that

$$\left\langle \frac{\delta H}{\delta \Phi} \right\rangle = \frac{\delta \langle H \rangle}{\delta \langle \Phi \rangle}\ , \text{ etc.} \tag{5.8}$$

Since there are no conservation laws constraining the variation of $\langle \Phi \rangle$ and of $\langle \pi \rangle$, we now derive from (5.7)

$$\delta \langle H \rangle = 0 = \int [- \langle \dot{\pi} \rangle\ \delta \Phi + \langle \dot{\Phi} \rangle\ \delta \pi] \tag{5.9}$$

and we conclude that

$$\langle \dot{\pi} \rangle = \langle \dot{\Phi} \rangle = 0. \tag{5.10}$$

Thus a neutral pion condensate wave function must be time independent.

Let $\langle \Phi(x) \rangle = \phi(r)$ and $\langle j(x) \rangle = J(r)$. Then the formally exact equation determining the condensate wave-function following either from (5.2) or (5.6) is

$$(\nabla^2 - m_\pi{}^2)\ \phi(r) - J(r) = 0. \tag{5.11}$$

We now imagine that all other field equations, namely (5.6) as well as those for the "nucleon" variables have been solved in terms of $\phi(r)$ and used to express J as a functional $J[\phi]$. We also assume that we are interested in values of $\phi(r)$ not too distant from the threshold of the condensation. If ϕ is a pseudoscalar, then J will be an odd functional of ϕ. Thus we write

$$J(r) = \int \Pi(r-r',0)\ \phi(r')$$

$$+ \frac{1}{3!} \int \Lambda(r-r',r'-r'',r''-r''')$$

$$\times\ \phi(r')\ \phi(r'')\ \phi(r''') + \dots\ , \qquad (5.12)$$

where, for instance,

$$\Pi(r-r',0) = \int d(t-t')\ \delta \langle j(x)\rangle/\delta \langle\Phi(x)\rangle\Big|_{\langle\Phi\rangle=0}$$

$$= \int d(t-t')\ \Pi(r-r',t-t'), \qquad (5.13)$$

and $\Pi(r,t)$ is the pion self-energy. The quantity Λ is the corresponding pion-pion scattering kernel. The quantities π,Λ, ... characterize the <u>normal fluid</u>.

Since for sufficiently small condensate density (5.11) and (5.12) must constitute the exact equation determining the condensate function, we can surmise that it must be the variational condition following from $W[\phi] = \langle H\rangle$ after all matrix elements other than ϕ have been eliminated by solution of the appropriate equations of motion. That is, Eqs. (5.11), (5.12) must be equivalent to the condition $\delta W[\phi]/\delta\phi = 0$. We conclude that the condensation energy has the form

$$W[\phi] = \frac{1}{2}\int d^3x[(\nabla\phi)^2 + m_\pi{}^2\phi^2] + \frac{1}{2}\int \phi(r)\,\Pi(r-r',0)\,\phi(r') + \frac{1}{4!}\int \phi(r)\,\phi(r')\,\phi(r'')\,\phi(r''') \times \Lambda(r-r',r'-r'',r''-r'''). \tag{5.14}$$

In this method, which is essentially that of Migdal[14], the problem "reduces" to obtaining convergent series for Π and for Λ. We shall comment in Sec. VI on the extent to which this has been done.

It is important in attempting to resolve the various controversies that have arisen, to remark that Migdal's method differs in principle from the mean field method as it has been applied by a number of authors.[17-21] To analyze this situation let us choose an example where

$$L^{(int)} = - j\Phi \tag{5.15}$$

and j is independent of Φ. With $\Phi = \phi + \Phi'$, $\langle\Phi'\rangle = 0$, we compute

$$W[\phi] = \frac{1}{2}\int [(\nabla\phi)^2 + m_\pi{}^2\phi^2] + \frac{1}{2}\int \langle[\pi'^2 + (\nabla\Phi')^2 + m_\pi{}^2\Phi'^2]\rangle + \int \langle j\rangle\phi + \int \langle j\Phi'\rangle + \langle H_N\rangle \tag{5.16}$$

as a formally exact expression. We observe first that the equations of motion for the <u>nucleons</u> are contained in the statement

$$\delta H_N + \int \Phi\delta j = \delta H_N + \int \delta(j\Phi) - \int j\delta\Phi = \int \sum_\alpha [-P_\alpha\delta Q_\alpha + \dot{Q}_\alpha\delta P_\alpha] = 0, \tag{5.17}$$

where Q_α, P_α are the canonical pairs for the nucleons. Consequently for the average in the ground state

$$\delta\langle H_N\rangle + \delta\int[\langle j\rangle\phi] + \delta\int\langle j\Phi'\rangle - \int\langle j\rangle\delta\Phi$$

$$= \delta\langle H_N\rangle + \int\phi\delta\langle j\rangle + \int\delta\langle j\Phi'\rangle = 0. \qquad (5.18)$$

Next we notice that in the mean field approximation we omit the second and fourth terms of (5.16), i.e.

$$W[\phi] = \frac{1}{2}\int[(\nabla\phi)^2 + m_\pi{}^2\phi^2] + \int\langle j\rangle\phi + \langle H_N\rangle. \qquad (5.19)$$

Variation of this expression with respect to ϕ yields in consequence of (5.18), the equation

$$(-\nabla^2 + m_\pi{}^2)\phi + J = \delta\int\langle j\Phi'\rangle/\delta\phi, \qquad (5.20)$$

which differs by the right hand side from the correct equation (5.11). This formal discrepancy can be rectified if we determine $\langle j\rangle$ from (5.18), omitting the last term. If this procedure is understood as part of the mean field approximation, we conclude that the latter is an approximate version of (5.14).

In practice then the method would consist in assuming a trial form for the condensate function

$\phi(r)$ for insertion into (5.14). With the further supposition that we have believable forms for Π and Λ, we then seek a minimum for $W[\phi]$ as a function of the parameters of $\phi(r)$ and of the nuclear density. Further comment on this case is delayed to Section VI. Any tenable model must yield Λ as a positive definite operator in order to insure stability of the condensate phase. For the neutral pion case the condensation energy can be determined without calculation of any observables relating to the nucleon components!

B. Charged Pion Condensate

This case involves subtleties not present for the neutral case. We take for granted the steps analogous to (5.1) - (5.4) and proceed immediately to the Hamiltonian

$$H = \int [\pi^\dagger \pi + \nabla\Phi^\dagger \cdot \nabla\Phi + m_\pi{}^2 \Phi^\dagger \Phi]$$

$$- \int L^{(\mathrm{int})} + \frac{\partial L^I}{\partial \dot{\Phi}^\dagger} \frac{\partial L^I}{\partial \dot{\Phi}} + H_N . \qquad (5.21)$$

where, e.g.

$$\pi(r,t) = \partial L / \partial \dot{\Phi}^\dagger (r,t) = \dot{\Phi} + \partial L^{(\mathrm{int})} / \partial \dot{\Phi} . \qquad (5.22)$$

The considerations leading to (5.10) must now

be altered. The variation of <H> is now subjected to the constraint of the various conservation laws. For example, imagine a system consisting of charged pions, neutrons and protons. Let

$$\rho_\pi = i(\Phi^\dagger \pi - \pi^\dagger \Phi) \tag{5.23}$$

be the pion charge density operator in units of the electron's charge and let ρ_p and ρ_n be the proton and neutron contribution to the baryon number operators. The variation of the energy is to be carried out subject to conservation of charge and of nucleons, i.e.

$$\delta[\langle H\rangle - \mu_\pi \int [\rho_\pi - \rho_p] - \mu_n \int (\rho_n + \rho_p)] = 0. \tag{5.24}$$

This equation has several important consequences. First, if we consider only the variation of the pion field, in place of (5.10) we obtain

$$\langle\dot{\Phi}\rangle = \dot{\phi}(r,t) = - i\mu_\pi \phi(r,t),$$

$$\langle\dot{\pi}\rangle = \dot{\pi}(r,t) = - i\mu_\pi \pi(r,t), \tag{5.25}$$

or, e.g.

$$\phi(r,t) = \exp[-i\mu_\pi t]\ \phi(r). \qquad (5.26)$$

Second, we must be clear that μ_π is the chemical potential for adding unit negative charge at equilibrium, which is not at all the same as adding a negative pion. Thus the equilibrium condensate will in general contain some relative proportions of $\pi_\pm$ and protons and μ_π is determined by addition of charge maintaining these proportions. Thus even in the limit of vanishing condensate $\mu_\pi \neq \mu_{\pi-}$ in general. Some confusion in the literature on this point has been settled by Sawyer.(19) Finally we see that μ_n is the chemical potential for adding a baryon (or a neutron) and

$$\mu_n = \mu_\pi + \mu_p. \qquad (5.27)$$

We now turn to the equations required to study condensate possibility. The analogue of (5.14) is now

$$W[\phi] = \int [|\nabla\phi|^2 + (m_\pi{}^2 - \mu_\pi{}^2)\ |\phi|^2]$$
$$+ \int [\phi^*(r)\ \Pi(r-r';\mu_\pi)\ \phi(r')]$$
$$+ \left(\frac{1}{2!}\right)^2 \int \phi^*(r)\ \phi^*(r')$$
$$\times\ \Lambda(r-r'', r'-r''', r-r')\phi(r'')\phi(r'''). \qquad (5.28)$$

The main change (in addition to obvious factors and definitions) is that frequency μ_π must replace frequency zero in both Π and Λ. The problem of the charged condensate is further complicated in that the dependence on μ_π and μ_n, implicit in (5.28) must be eliminated through the specification of charge density ρ_Q and baryon density ρ_B. This means computation of

$$<[\rho_\pi - \rho_p]> = \rho_Q,$$

$$<[\rho_p + \rho_N]> = \rho_B, \qquad (5.29)$$

as part of the minimization procedure. As will be seen in the next section, these quantities can also profitably be treated as functional series in the condensate wave function.

VI. SUMMARY AND CRITIQUE OF CALCULATIONS ON POSSIBLE PION CONDENSATION IN INFINITE MATTER

A. Crude Model for Pion Condensation in Neutron Matter

We shall first show, using the method of Migdal consistently that π^- condensation can occur in a highly simplified model. Our result will agree with that given by Sawyer[19], using another method.

The transition point is determined by the

solution of the equation

$$k^2 + m_\pi^2 - \mu_\pi^2 + \Pi(k,\mu_\pi) = 0. \qquad (6.1)$$

We treat Π to lowest order perturbation theory in the fixed source limit. Below it is shown that this is equivalent to the mean field approximation in the same limit. The kernel Π is calculated in a familiar way from the limit of the simple closed loop diagram containing a neutron hole and a proton particle. The result is

$$\Pi(k,\omega) = -\frac{2f^2k^2\rho}{\omega m_\pi^2} \equiv -\frac{2M_k^2\rho}{\omega}. \qquad (6.2)$$

Here $\rho = \rho_B$ is the baryon density and $f^2 = 1.1$.

To utilize (6.1) and (6.2) we need a relation between ρ and μ_π. This is obtained from the first of Eqs. (5.29) for $\rho_Q = 0$, namely

$$\langle\rho_\pi\rangle = \rho_p. \qquad (6.3)$$

Consider a plane wave condensate in the z direction

$$\langle\Phi(r)\rangle = \phi \exp(ikz)$$

$$\langle\pi\rangle = \langle\dot{\Phi}\rangle = i\mu_\pi \phi \exp(ikz), \qquad (6.4)$$

We thus have

$$\langle \rho_\pi \rangle = 2\mu_\pi |\phi|^2. \tag{6.5}$$

To calculate ρ_p, the correct Hamiltonian to use in the mean field approximation is the sum of the pion part and of an interaction of form

$$H^{(int)}(t) = -\sqrt{2}\,[iM_k\; \psi_n^\dagger(t)\sigma_3\; \psi_p(t)\phi + h.c.]. \tag{6.6}$$

Here $\{\psi_p, \psi_p{}^\dagger\} = 1$, etc. and

$$\langle \psi_n^\dagger\; \psi_n \rangle = \rho_n,\; \langle \psi_p^\dagger\; \psi_p \rangle = \rho_p. \tag{6.7}$$

These are boundary values of the Green's function

$$G^{(ab)}(t) = i\; \langle T(\psi_a(t)\psi_b^\dagger(0))\rangle. \tag{6.8}$$

Thus

$$\rho_p = iG^{(pp)}(t\to 0_-) = iG^{(pp)}(0)$$

$$= \frac{i}{2\pi}\oint_{UHP} d\omega \tilde{G}^{(pp)}(\omega), \tag{6.9}$$

where

$$\tilde{G}^{(ab)}(\omega) = \int e^{i\omega t} G^{(ab)}(t), \qquad (6.10)$$

and the contour is closed in the upper half plane.

Including terms $[-\mu_n\psi_n^\dagger\psi_n - \mu_p\psi_p^\dagger\psi_p]$ in the Hamiltonian, standard methods yield the following equations

$$(\partial_t - i\mu_p)\ G^{(pp)}(t) = i\ \delta(t) + \sqrt{2}\ M_k\ \sigma_3\phi^* G^{(np)}(t)$$

$$(\partial_t - i\mu_n)\ G^{(np)}(t) = -\sqrt{2}\ M_k\sigma_3\phi\ G^{(pp)}(t). \qquad (6.11)$$

Utilizing (6.10), we obtain

$$\tilde{G}^{(pp)}(\omega) = \frac{-i}{\omega + \mu_n}\sqrt{2}\ M_k\phi\ \sigma_3\ \tilde{G}^{(pp)}(\omega)$$

$$= i\ \tilde{G}_o^{(nn)}(\omega)\ \sqrt{2}\ M_k\phi\sigma_3\ \tilde{G}^{(pp)}(\omega), \qquad (6.12)$$

where

$$[\tilde{G}_o^{(ab)}(\omega)]^{-1} = -(\omega+\mu_a)\ \delta_{ab}, \qquad (6.13)$$

and

$$\tilde{G}^{(pp)}(\omega) = \tilde{G}_o^{(pp)}(\omega) + \tilde{G}_o^{(pp)}(\omega)\ \tilde{G}_o^{(nn)}(\omega)$$

$$\times\ 2M_k^2\ |\phi|^2\ \tilde{G}^{(pp)}(\omega). \qquad (6.14)$$

Equation (6.14) is easily solved exactly, but comparing (6.3), (6.5), (6.9) and (6.14), it is seen to be sufficient to replace $\tilde{G}^{(pp)}$ by $\tilde{G}_o^{(pp)}$ in the second term of (6.14). We can then calculate ρ_p from (6.9). To do this we need the representations

$$\tilde{G}_o^{(pp)}(\omega) = \frac{-\rho_p}{\omega + \mu_p - i\eta} - \frac{(1-\rho_p)}{\omega + \mu_p + i\eta},$$

$$\approx \frac{-1}{\omega + \mu_p + i\eta}, \quad \rho_p \to 0, \tag{6.15}$$

$$\tilde{G}_o^{(nn)}(\omega) = \frac{-\rho}{\omega + \mu_n - i\eta} - \frac{(1-\rho)}{\omega + \mu_n + i\eta} . \tag{6.16}$$

We then obtain ($\mu_\pi = \mu_n - \mu_p$) from (6.9)

$$\rho_p = \frac{2M_k^2 |\phi|^2}{\mu_\pi^2} \rho, \tag{6.17}$$

or remembering (6.3) and (6.5), we find

$$\mu_\rho^3 = \rho M_k^2. \tag{6.18}$$

With (6.1) and (6.2), (6.18) yields

$$\omega = \sqrt{3}\, \mu_\pi, \tag{6.19}$$

$$\rho = \omega^3/M_k{}^2 \, (3)^{3/2}. \qquad (6.20)$$

The minimization of (6.20) now gives

$$k_c = \sqrt{2}\, m_\pi, \qquad (6.21)$$

$$\rho_c = m_\pi{}^3/2f^2. \qquad (6.22)$$

The specific numbers here are of little interest, because of the crudity of the model. The occurrence of a phase transition in this simplified model is of great interest as a stimulus for more careful and detailed investigation. It is important to emphasize that the transition should be characterized as a charged pion phase transition. The dynamics will determine the ratio (m_{π_+}/m_{π_-}). For a plane wave mode, it is simple to show[21] that

$$n_{\pi_+}/n_{\pi_-} = (\omega-\mu_\pi)^2/(\omega+\mu_\pi)^2. \qquad (6.23)$$

As one more formal result we shall show that in the approximation employed, Migdal's method and the mean field method are the same. In this connection (6.18) is already the mean field result (as well as the perturbation result). It remains to show that the minimization of the mean field energy at threshold agrees with (6.1) and (6.2).

We have first

$$<H_\pi> = <\int [\pi^\dagger \pi + \nabla\Phi^* \nabla\Phi + m_\pi^{\;2} \;\Phi^\dagger \Phi]>$$

$$= [\mu_\pi^{\;2} + k^2 + m_\pi^{\;2}] \; |\phi|^2. \qquad (6.24)$$

We must now calculate the expectation value of (6.6). Toward this end we need the solution (6.14). We find that $G^{(pp)}(\omega)$ has two poles[21]

$$E_\pm = - \frac{1}{2} (\mu_p + \mu_n) \pm [\mu_\pi^{\;2} + \delta M^2 \; |\phi|^2]^{1/2}, \qquad (6.25)$$

and that boundary conditions can be incorporated by writing

$$\tilde{G}^{(pp)}(\omega) = - \frac{(\omega + \mu_n)}{(E_+ - E_-)} \left[\frac{\rho_+}{\omega - E_+ - i\delta} + \frac{(1-\rho_+)}{\omega - E_+ - i\delta} + (+ \leftrightarrow -)\right], \qquad (6.26)$$

where $\rho_\pm$ will be determined imminently (with the equal help of the easily determined corresponding neutron Green's function

$$\tilde{G}^{(nn)}(\omega) = -\frac{(\omega+\mu_p)}{(E_+-E_-)} [\quad], \qquad (6.27)$$

where the bracket is the same as in (6.26)).

The application of (6.9) and the corresponding equation for ρ_n leads upon inversion to the following determination

$$\mu_\pi \rho_\pm = -(E_\mp+\mu_p)\,\rho_p + (E_\mp+\mu_n)\,\rho_n. \qquad (6.28)$$

With the aid of the analogue of (6.9), (6.12), (6.26) and (6.28), we now calculate

$$\langle\psi_p^\dagger \sigma_3 \psi_n\rangle = -i\sqrt{2}\,M_k\phi\,[\rho_+ - \rho_-]/[E_+ - E_-]$$

$$= i\sqrt{2}\,M_k\phi\,[\rho - 2\rho_p]/\mu_\pi, \qquad (6.29)$$

which we shall utilize in the limit $\rho_p \to 0$. This means that to terms quadratic in $|\phi|^2$ the mean field energy becomes

$$W[\phi] = \langle H\rangle = [\mu_\pi^2 + k^2 + m_\pi^2 - \frac{4M_k^2\rho}{\mu_\pi}]\,|\phi|^2. \qquad (6.30)$$

The square bracket is, however, the same as (6.1) and (6.2) if we remember (6.18). Since the condition for transition is $\overline{W} = 0$, we have reached the result sought.

In contrast to the charged pion case, the neutral pion case does not have a sensible fixed nucleon limit.[19] However, mean field theory, including the nucleon kinetic energy, would again appear to yield a possible second order phase transition.

B. Present Status and Possible Future Program

It requires unusual optimism to accept uncritically the conclusions of any calculations of pion condensation carried out to date. For instance, there is no calculation which takes consistently into account at the same time all of the following effects present in the pion self-energy operator. (i) Resonant pion-nucleon scattering, (ii) Multiple pion-nucleon scattering, (iii) Short range nuclear correlations, (iv) Pion-pion interaction.

Regarding the problem from the point of view of Migdal's method, the main task is the calculation of the pion-self energy -- quite analogous to the problem of computing the pion-optical potential in nuclear matter. It is our opinion that if the interesting range of densities does not much exceed nuclear densities, then the most promising approach to the problem is a nucleon hole line expansion, analogous to the Brueckner-Goldstone expansion for nuclear matter. A start has been made on this program [30-33], but much work remains to be done. Similar methods [34] should suffice for the calculation of the nucleon densities needed for the study of the conservation laws of charge, baryon number, etc. We hope to return to these matters in a future publication.

REFERENCES

1. W. Peiper and W. Greiner, Z. Physik 218 (1969) 327.
2. B. Muller, H. Peitz, J. Rafelski and W. Greiner, Phys. Rev. Letters 28 (1972) 1235.
3. B. Müller, J. Rafelski and W. Greiner, Z. Physik 257 (1972) 62.
4. B. Müller, J. Rafelski, and W. Greiner, Z. Physik 257 (1972) 183.
5. H. Peitz, B. Müller, J. Rafelski and W. Greiner, Lett. Nuovo Cimento 8 (1973) 37. Preprint, U. of Frankfurt, 1973.
6. J. Rafelski, B. Müller, W. Greiner, The Charged Vacuum in Overcritical Fields, to be published in Nuclear Physics.
7. V. S. Popov, Yad. Fiz 12 (1970 429 [Translation: Sov. J. Of Nucl. Phys. 12 (1971) 235].
8. V. S. Popov, Zh. Eksp. Teor. Fiz. 59 (1970) 965 [Translation: Soviet Physics JETP 32 (1972 526].
9. Ya. B. Zel'dovich and V. S. Popov, Usp. Fiz. Nauk 105, 3 (1971)[Sov. Phys. Usp. 14, 6 (1972)].
10. L. Fulcher and A. Klein, Phys. Rev. D8 (1973) 2455.
11. L. Fulcher and A. Klein, to be published in Annals of Physics.

12. A. B. Migdal, Zh. Eksp. Teor. Fiz. 61 (1971) 2209 [Sov. Phys. JETP 34 (1972) 1184]. Our treatment differs in a number of technical aspects with the work of this reference.
13. A. B. Migdal, Physics Letters 45B (1973) 448.
14. A. B. Migdal, Phys. Rev. Letters 31 (1973) 257.
15. A. B. Migdal, Nucl. Phys. A210 (1973) 421.
16. A. B. Migdal, Phys. Letters 47B (1973) 96.
17. R. F. Sawyer and D. J. Scalapino, Phys. Rev. D7 (1973) 953.
18. R. F. Sawyer and A. C. Yao, Phys. Rev. D7 (1973) 1579.
19. R. F. Sawyer, Phys. Rev. Letters 31 (1973) 1556.
20. G. Baym, Phys. Rev. Letters 30 (1973) 1340.
21. G. Baym and E. Flowers, submitted to Nuclear Physics.
22. S. Barshay, G. Vagradov, and G. E. Brown, Phys. Letters 43B (1973) 359.
23. S. Barshay and G. E. Brown, Phys. Letters 47B (1973) 107.
24. V. Fano, Phys. Rev. 124 (1961) 1866.
25. M. L. Goldberger and K. M. Watson, Collision Theory (John Wiley and Sons, Inc., New York, 1964) Chap. 8.
26. E. H. Wichmann and N. M. Kroll, Phys. Rev. 101 (1956) 843.
27. H. Feshbach and F. Villars, Rev. Mod. Phys. 30 (1958) 24. Our representation differs by

rotation in τ space from the one given in this paper.

28. H. Snyder and J. Weinberg, Phys. Rev. 57 (1940) 307.
29. L. I. Schiff, H. Snyder and J. Weinberg, Phys. Rev. 57 (1940) 315.
30. C. B. Dover, J. Hüfner, and R. H. Lemmer, Ann. Phys. (N.Y.) 66 (1971) 248.
31. C. B. Dover, Ann. Phys. (N.Y.) 79 (1973) 441.
32. H. A. Bethe, Phys. Rev. Lett. 30 (1973) 105.
33. J. M. Eisenberg and H. J. Weber, Phys. Lett. 45B (1973) 110.
34. J. Hüfner and C. Mahaux, Ann. Phys. 73 (1972), 525.

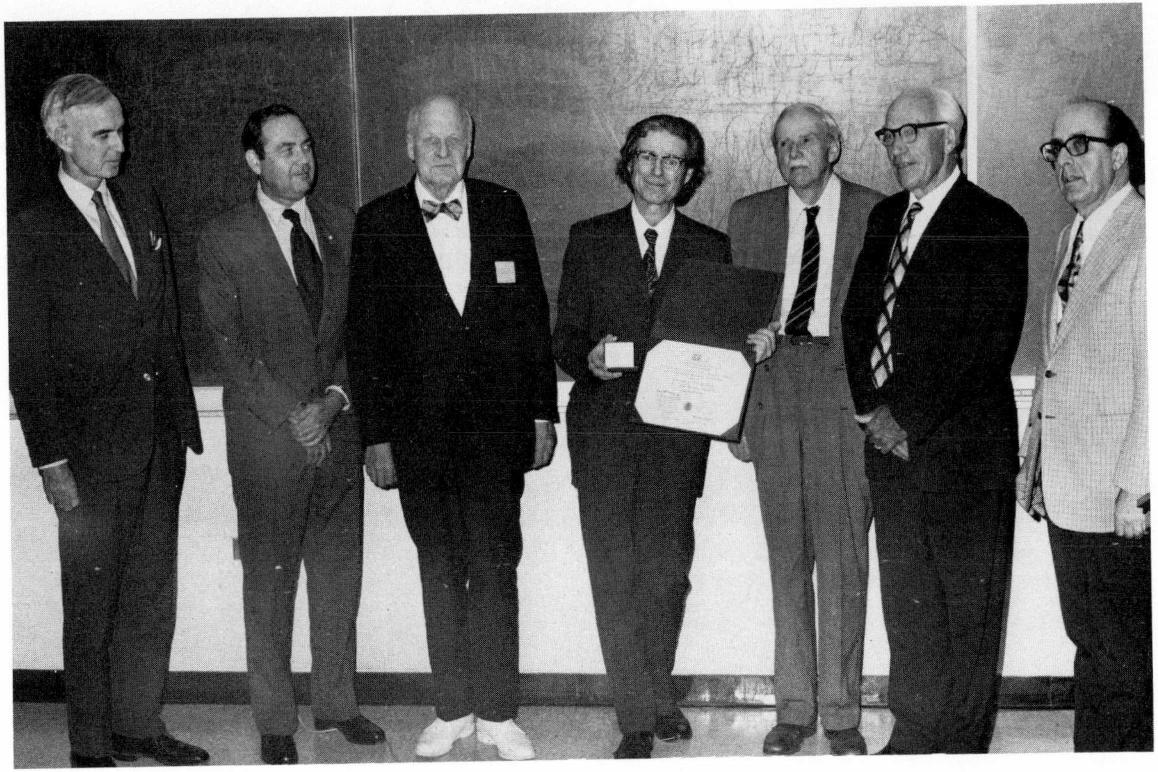

Sixth Annual J. Robert Oppenheimer Memorial Prize, January 9, 1974. (From Left: Willis E. Lamb, Jr.; David Blumberg; Lars Onsager; Edwin E. Salpeter, Prize Recipient; P.A.M. Dirac; John Eccles; and Behram Kursunoglu.)

DYING STARS AND REBORN DUST

E. E. Salpeter

Laboratory of Nuclear Studies

Cornell University

I am happy to help honor the memory of J. Robert Oppenheimer at this occasion. I started my scientific career as a quantum field theorist and, on numerous pilgrimages to the Princeton Institute, I was inspired and helped by Oppie's enthusiasm and insight. Although Oppenheimer is not so well known in astrophysical circles, he and his associates actually did much of the pioneering work on the gravitational collapse and structure of neutron stars long before any were discovered. Neutron stars lie at the heart of supernova remnants which I will discuss at the end of this report. My topic deals with one aspect of the symbiotic relationship between stars on the one hand and interstellar gas and dust on the other -- the effect of dying stars (more specifically of planetary nebulae and of supernovae) on producing and replenishing dust-grains. I shall mainly review the current status but end with a speculative conjecture.

In spiral galaxies like our own, interstellar dust occurs together with interstellar gas in a disk a few hundred light-years thick. Some information on the optical properties of these dust grains has been available for a long time from studies in the visible spectrum of reflection nebulae and of the extinction and polarization curves for light from distant stars. This information has been greatly augmented in the last decade or so by infrared observations and by the two OAO satellites working in the ultraviolet. Detailed reviews were presented by Greenberg (1973) and by Woolf (1973) at a recent Symposium and some of the gross features are shown schematically in Fig. 1, where the extinction efficiency $Q(\lambda)$ at wavelength λ (multiplied by λ) is plotted against λ. The two "humps" in the infrared are generally agreed to be absorption features common to most types of silicate grains and the peak near $\lambda \sim 2{,}500$ Å is thought by some to be due to small graphite grains, but this assignment is still controversial. The most prominent interstellar dust-grains must have radii a of order 0.03μ to 0.2μ, but the further rise of $\lambda(Q)\lambda$ in the far ultraviolet indicates the presence of some very small grains with $a \lesssim 100$ Å.

If one assumes some specific model of the distribution of sizes and chemical composition of the interstellar grain, which fits the <u>shape</u> of the observed extinction (and polarization) curves,

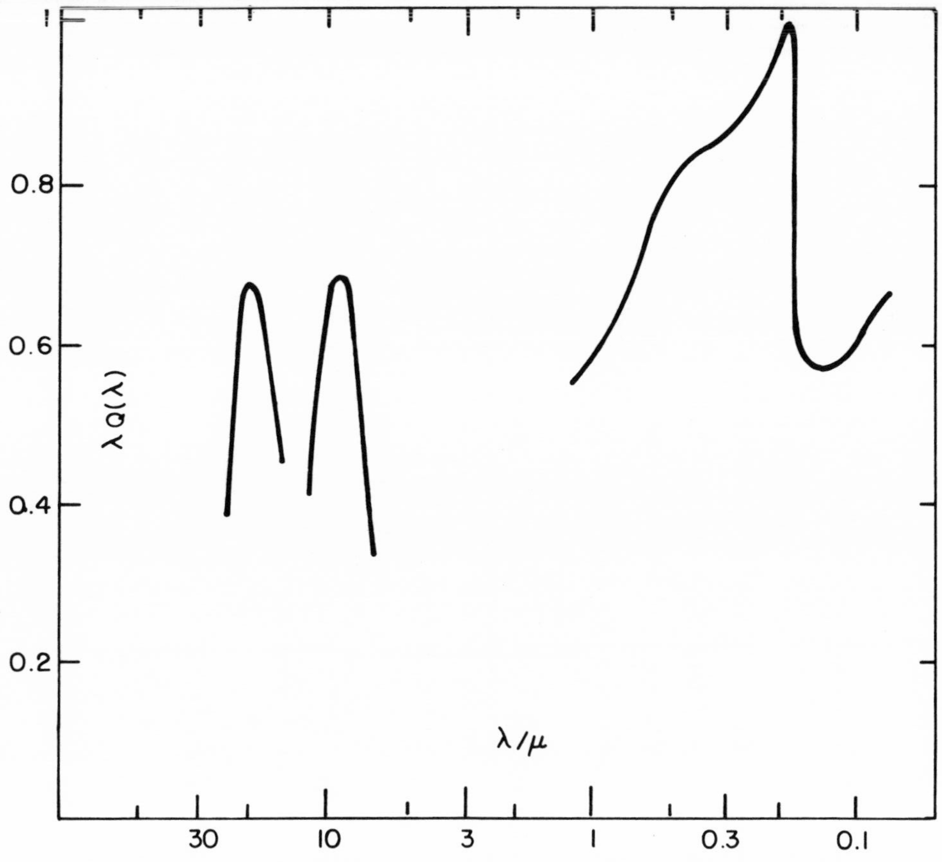

Figure 1: The extinction efficiency $Q(\lambda)$ at wavelength λ (multiplied by λ) of typical interstellar dust-grains plotted (in arbitrary units) against λ.

<u>Table 1.</u> The "cosmic abundance" A_c, by numbers of atoms relative to hydrogen, for a few elements and the ratio of the abundance A_{gas} in the interstellar gas to A_c.

	C	O	Mg	Al + Ca
$\log_{10} A_c$	-3.5	-3.2	-4.5	-5.5
$\log_{10} \frac{A_{gas}}{A_c}$	-1.2	~-0.5	-1.2	-(2.5 to 4)

one can calculate (Greenberg, 1974) the total amount of grain material required to fit the magnitude of the extinction. Without making specific model assumptions, one can still derive lower limits to the required amount of grain material from the Kramers-Kronig dispersion relation (see Purcell and Pennypacker 1973; Caroff et. al. 1973). It is of interest to compare the conjectured amounts of grain material with the amounts in grains and gas combined given by the so-called "cosmic abundances" (relative to hydrogen, whose density in interstellar space is known fairly well from 21 cm observations). These "cosmic abundances," obtained mainly from the solar atmosphere where there are no grains, are given in the first line of Table 1 for a few of the most relevant elements (nitrogen is slightly less abundant than carbon; silicon has comparable abundance to magnesium, which is the most abundant metal to form silicates). Analyses of the extinction data show that much of the Mg and Si in interstellar space must be in the form of grains (if the infrared "humps" are indeed due to silicates). Furthermore, there must be more material in addition to silicates in the grains (to fit the ultraviolet data). Most of the additional material is thought to be in compounds derivable from the "ices" of abundant elements, CH_4, NH_3 and H_2O. The full amount of these ices obtainable from the cosmic abundances is not absolutely required, if most grains are near the optimum size

(Greenberg 1974). However, we have no direct evidence on how much C, N and O is locked up in very small (a < 50 Å) or very large (a > 0.5μ) grains whose optical efficiency is low.

To summarize present views on the chemical composition and abundance of interstellar dust-grains (see also Aannestad and Purcell 1973). There is plenty of observational data and it is clearcut whether a given theoretical model fits the data or not. However, there is not sufficient observational data to be able to derive a unique model from it. Much of the condensable material (such as the metals, C, Si, etc.) in interstellar space must be in the form of grains, but the data does not specify whether "much" means 25% or 100%.

The amount of various elements in the form of gas atoms (or of simple molecules) in interstellar space can be measured by means of optical absorption lines and compared with the amount of hydrogen. Much of this data comes from UV observations with the Copernicus satellite (Morton et al 1973, Spitzer et al 1973), but ground-based observations are also involved (White 1974). The relevant data and its significance was summarized recently by Field (1973, 1974) and a few examples are given in the second line of our Table 1. The results are not yet quantitatively reliable, but suggest the following rather surprising trends: (i) Many elements are slightly underabundant in the interstellar gas (compared with cosmic abundances), such as O; (ii) a few, including C and Mg (also

Mn and Fe) are underabundant by slightly more than a factor of ten and (iii) especially Ca and probably also Al and Ti are underabundant by a factor of a thousand or more!

Before discussing the possible significance of the abundance data, we need the thermochemical information summarized schematically in Fig. 2. To form grains from the vapor phase we need temperatures in the range of 1000 to 2000°K. Above the dashed line labelled CO in Fig. 2, carbon monoxide is extremely stable in gaseous form. Therefore, whichever of C and O has the lower abundance (by number of atoms) is locked in the gas-phase as CO and only the more abundant partner is free to form grains. Usually (see the cosmic abundances in Table 1) O is slightly more abundant than C and the most refractory solids which can be built from oxygen-rich material are the various silicates. As indicated in Fig. 2, minerals involving calcium and aluminum silicates condense out at the highest temperature. As material is cooled, these minerals will condense out first to form very small grains by homogeneous nucleation; upon further cooling the more abundant magnesium silicates will condense using these very small grains as condensation nuclei (Salpeter 1974). In a small (but not negligible) fraction of stars carbon is more abundant than oxygen and the most important grains formed are carbon-rich particles with a structure similar to

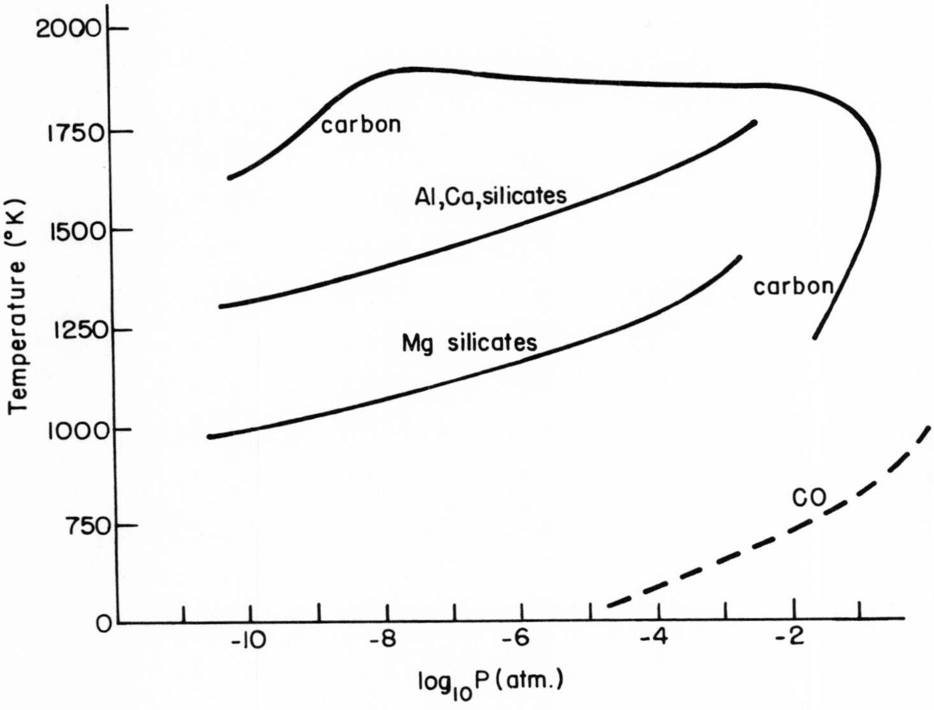

Figure 2: A schematic pressure-temperature phase-diagram: Carbon monoxide is stable above the dashed curve; for carbon-rich material graphite-like grains condense out near the curve labelled "carbon"; for oxygen-rich material various types of silicates condense out between the two curves labeled "silicates".

graphite but probably with a small admixture of hydrogen.

The possibility of making dust grains in the atmospheres of cool stars was already raised some time ago by Hoyle and Wickramsinghe (1962) and by Donn and Stecher (1965). The situation is particularly simple if the star's effective surface temperature T_e is close to or even below the condensation temperatures for the solids ($T_e < 2000^oK$, say). If the star's luminosity L is sufficiently large so that radiation pressure acting on the grains exceeds the gravitational force on grains and gas (H and He) combined ($L >> 10^3 L_\odot$, say), grains are not only made but also injected into interstellar space by the resultant outflow. Such large values of L and small values of T_e require very large values of stellar radius R, reached only by the coolest supergiants (especially near the greatest expansion of long-period variable stars, e.g. Mira variables). The situation is particularly favorable for the rarer carbon-rich stars, since graphite-like carbon particles condense at relatively high temperature and have a high opacity which makes radiation pressure more important (Fix 1969). The theory is more complicated and uncertain for the production of silicate grains in oxygen-rich stars (Kamijo and de Jong 1973), but fortunately there is some observational evidence from infrared emission features in the spectra of cool stars for the presence of silicate grains (Jennings and Dyck 1972, Cohen and Gaustad 1973, Woolf 1973). The total outflow rate of grains from such cool supergiants is not clear yet and other sources of grain production are also of interest. Very young and very massive stars (Hoyle, Solomon and Woolf 1973) may also contribute, but stars of moderate mass (a few solar masses, say) are probably most important. During some late stages of their evolution a copious outflow of gas takes place, which is not caused by grains but grains can form subsequently as the outflowing stream of gas cools.

Only a moderate amount of mass-loss is thought to occur when a star with a helium-core first evolves away from the main sequence into the red giant region. The situation is more favorable during the later stages of evolution, when a star has a small core rich in C, O and Ne, a helium-rich mantle and a hydrogen-rich outer envelope. This envelope expands, the luminosity is quite large ($L \gtrsim 10^4 L_\odot$) and at some stage the envelope becomes unstable to hydrodynamic outflow. This outflow is the basis of the formation of a planetary nebula, where a mass of the order of $\sim 1 M_\odot$ is ejected over a period of about a thousand years. Recent observational data (Cahn and Kaler 1971) on planetary nebulae suggest that a large fraction ($\gtrsim \frac{1}{2}$) of all "star-deaths" occur only after the star passes through this phase. The stellar remnant in such cases is a white dwarf, contrasted with the formation of a neutron star during a supernova event. Supernovae occur less frequently than planetary nebulae by a factor of about 100; we shall nevertheless return to a conjecture about the effect of an expanding supernova shell on the surrounding interstellar medium.

During the ejection of a planetary nebula shell the material has to flow out a few stellar radii before it is sufficiently cool for all the silicates to condense out. The gas density is rather low (and not well known theoretically) at this stage and purely theoretical calculations of the efficiency of grain formation are not yet very reliable (Salpeter 1973). Fortunately, direct observational evidence on infrared emission from planetary nebulae indicates the presence of radiating dust-grains (probably silicates) more than 10^4 years after the ejection of the nebula shell (Woolf 1973, Leibowitz 1973). Planetary nebulae are probably one of a small number of major sources for interstellar silicate grains and current estimates for the total production rates of such grains are encouraging. The total amount of material processed through magnesium silicate grains, for instance, in the lifetime of our galaxy ($\sim 10^{10}$ years) may even be a few times larger than the present amount of Mg and Si in interstellar space. Such an inequality is possible because more than 90% of the total mass of our galaxy has been processed from interstellar material into stars and many grains formed earlier have been used up in star-formation.

Estimates for grain production in stellar atmospheres and in planetary nebulae indicate that a fairly large (but <u>not</u> overwhelmingly large) fraction of the interstellar Mg, Si, Ca and Al, say, should be in the form of grains. This would predict a small degree of underabundance of these elements in the interstellar <u>gas</u> but is incompatible with the very large degree of underabundance observed for Ca and Al (see Table 1). Furthermore, the <u>moderately</u> large underabundances in the gas of C and O are also incompatible (Greenberg 1974, Field 1974) with the assumption that grains are grown only in stellar atmospheres and planetary nebulae without further reprocessing. As mentioned, a substantial fraction of the C and O in the outflowing material is in the form of gaseous CO which does not condense easily (and is later photodissociated into atomic C and O); more generally, the various "ices" cannot be made in this manner. In fact, planetary nebulae older than about 10^4 years provide a direct illustration of the dilemma. The nebula shell is optically thin to ultraviolet radiation from the central star of the nebula; theoretically, only refractory materials like silicates (or graphite) can survive in solid form, but not the more volatile "ices" made from C, N and O. Observationally, we have direct evidence (Peimbert 1971) for the presence of nitrogen and oxygen in gaseous atomic form in these planetary nebula shells which eventually must get mixed in with the interstellar gas. One also has to contend with an additional difficulty. In some regions of interstellar space conditions are such that sputtering will partially destroy grains, which should also feed more atoms into the interstellar gas (Aannestad 1973).

Of various ways out of this dilemma, one is connected with star-birth and another with star-death. Stars are often born in a group out of a massive, dense interstellar cloud of gas and dust. Only a small fraction of the cloud's mass goes into stars and the rest of the cloud is re-expanded by the radiation from the newly-formed stars. It has been suggested (Field 1974) that such cloud-material just before its re-expansion may be a favorable site for replenishing silicate (or graphite or iron) grains and for growing ice-mantles on them. It is hard to investigate this suggestion quantitatively in our present state of ignorance on the details of star formation. High gas-density favors and high stellar luminosity hinders the process and we do not know the density-temperature history sufficiently well. We finally turn to another tentative conjecture which involves one form of star-death, namely supernovae -- or, more correctly, the effect of the expanding supernova shell more than 10^5 years after the actual explosion.

Supernova explosions occur about one hundred times less frequently than planetary nebulae and

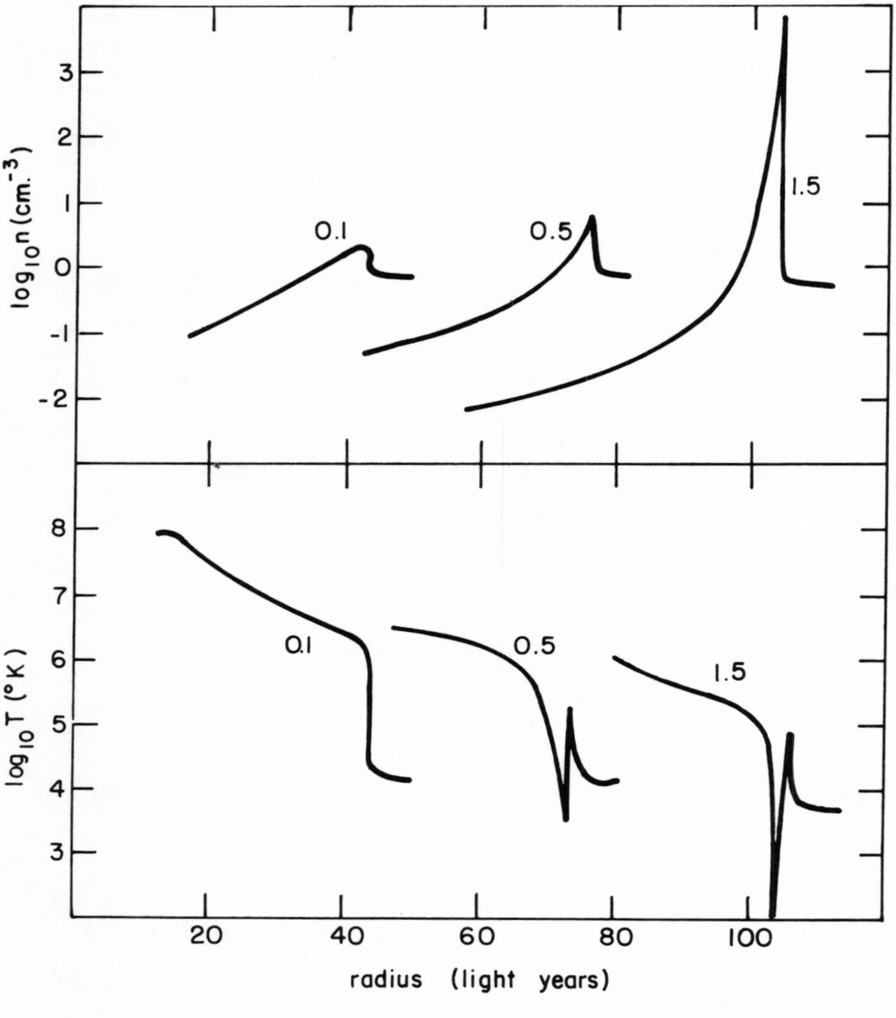

<u>Figure 3</u>: Theoretical blast-wave results (with radiative losses) for density n (in H-atoms cm^{-3}) and temperature T (in ^{o}K) plotted against radius for various times t after a supernova explosion. An explosion energy ~ 5 x 10^{50} ergs and an interstellar medium with n ~ 0.5 was assumed. Curves are labelled with the value of t in units of 10^5 years.

the ejected shell mass per event is comparable (~1 $M_\odot$). The mass ejected in the shell is of little direct importance, but the kinetic energy in the explosion is about 10^4 times larger than in a planetary nebula and the explosion causes a blast-wave to propagate out through the interstellar medium. A number of analytic studies have been made (1972), as well as numerical calculations (Chevalier 1974, Straka 1974, Mansfield and Salpeter 1974) to follw the progress of the blast-wave, including the effects of hydrodynamics and of radiative cooling. At approximate pressure-equilibrium, radiative cooling is faster at <u>lower</u> temperatures (at least for $T \gg 10^4$ °K) and this results in a "thermal instability". After less than 10^5 years a thin shell forms near the outer edge of the blast wave, which gets cooler, denser and thinner (but <u>more</u> massive) as time progresses (somewhat like a snow-plow plus a compressor). The density and temperature profiles at various time epochs are shown in Fig. 3, schematically, for a typical core. Note in particular the fairly hot shock-region, just outside the cool shell, through which the interstellar gas and dust must pass before becoming incorporated into the dense shell. Some sputtering of grains may take place in this shock-region, but this becomes unimportant when the shock-temperature drops well below 10^5 °K (the grain temperature is orders of magnitude lower). We then have a rather long period (of order 10^6 years) when interstellar gas and dust is compressed to rather high densities (up to ~ 10^4 H-atoms/cm^3, say), mostly at temperatures of $\lesssim 100^\circ K$. This condition may be quite favorable for condensing out more material from the interstellar gas (remaining metals as well as various "ices"). Unfortunately this conjecture is still quite controversial. Although an appreciable fraction of interstellar matter may be in such thin, dense shells, the theoretical uncertainties are great for such old supernova remnants due to uncertain effects of magnetic pressure and turbulence. It is also difficult to observe such shells optically, because they are so thin, but it may nevertheless prove possible to detect them in the form of thin rings projected against the sky where dust absorption makes background stars appear faint (a kind of "negative stellar ring").

The author is the J. G. White Professor of Physical Sciences at Cornell University, Ithaca, New York. This paper is based in part on a talk delivered at the J. Robert Oppenheimer Memorial Prize Ceremonies on January 9th, 1974, at the Center for Theoretical Studies, University of Miami, Coral Gables, Florida.

REFERENCES

Aannestad, P.A. 1973, Astrophys. J. Suppl. 25 (No. 217), 205.

Aannestad, P.A. and Purcell, E.M. 1973, Ann. Rev. Astron. and Astrophys. 11, 309.

Cahn, J.H. and Kaler, J.B. 1971, Astrophys. J. Suppl. 22 (No. 189), 319 .

Caroff, L., Petrosian, V., Salpeter, E., Wagoner R., and Werner, M. 1973, M.N.R.A.S. 164, 295.

Chevalier, R.A. 1974, Astrophys. J. (in print).

Cohen, M. and Gaustad, J.E. 1973, Astrophys. J. 186, L131.

Cox, D.P. 1972, Astrophys. J. 178, 143, 159 and 1969.

Donn, B.D. and Stecher, T.P. 1965, Astrophys. J. 142, 1681.

Field, G. B. 1973, paper at "The Dusty Universe" Symposium, Cambridge, Mass.

Field, G.B. 1974, Astrophys. J. 187, 453.

Fix, J. D. 1969, M.N.R.A.S. 146, 37 and 51.

Greenberg, J. M. 1973, paper at "The Dusty Universe" Symposium, Cambridge, Mass.

Greenberg, J. M. 1974 (unpublished).

Hoyle, F. and Wickramsinghe, N. C. 1962, M.N.R.A.S. 124, 417.

Hoyle, F., Solomon, P. and Woolf, N. 1973, Astrophys. J. 185, L89.

Jennings, M. C. and Dyck, H. M. 1972, Astrophys. J. 177, 427.

Kamijo, F. and de Jong, T. 1973, Astron. and Astrophys. 25, 371.

Leibowitz, E. M. 1973, Astrophys. J. 186, 899.

Mansfield, V. N. and Salpeter E. E. 1974, Astrophys. J. (in print).

Morton, D., Drake, J., Jenkins, E., Rogerson, J., Spitzer, L., and York, D. 1973, Astrophys. J. 181, L103.

Peimbert, M. and Torres-Peimbert, S. 1971, Astrophys. J. 168, 413.

Purcell, E. M. and Pennypacker, C. R. 1973, Astrophys. J. 186, 705.

Salpeter, E. E. 1973, J. Chem. Phys. 58, 4331, and paper at "The Dusty Universe" Symposium, Cambridge, Mass.

Spitzer, L., Drake, J., Jenkins, E., Morton, E., Rogerson, J. and York, D. 1973, Astrophys. J. 181, L103.

Straka, W.C. 1974, Astrophys. J. (in print).

White, R. E. 1974, Astrophys. J. 187, 449.

LIST OF PARTICIPANTS

Edward Ames
Department of Economics
State University of
New York
at Stony Brook

Joseph Aschheim
Department of Economics
George Washington
University

M.A.B. Beg
Department of Physics
Rockefeller University

Robert Blumenthal
Department of Health,
Education and Welfare
National Institutes of
Health

Martin Bronfenbrenner
Department of Economics
Duke University

E.R. Caianiello
Consiglio Nazionale delle
Ricerche
Laboratorio di Cibernetica

Mou-Shan Chen
Center for Theoretical
Studies
University of Miami

Anthony Colleraine
Department of Physics
Florida State University

P.A.M. Dirac
Department of Physics
Florida State University

John Eccles
Department of Physiology
State University of
New York
at Buffalo

Erich A. Farber
Department of Mechanical
Engineering
University of Florida

Sidney Fox
Institute for Molecular
and Cellular Evolution
University of Miami

Nicholas Georgescu-Roegen
Department of Economics
Vanderbilt University

Donald A. Glaser
Department of Molecular
Biology
University of California
at Berkeley

Melvin Gottlieb
Plasma Physics Laboratory
Princeton University

Gary Higgins
Lawrence Livermore
Laboratory
University of California

Joseph Hubbard
Center for Theoretical Studies
University of Miami

C.S. Hui
Center for Theoretical Studies
University of Miami

Henry Hurwitz
General Electric Company
Schenectady, New York

Abraham Klein
Department of Physics
University of Pennsylvania

Behram Kursunoglu
Center for Theoretical Studies
University of Miami

Willis E. Lamb, Jr.
Physics Department
Yale University

Joseph Lannutti
Department of Physics
Florida State University

Sydney Meshkov
Radiation Theory Section
National Bureau of Standards

Stephan L. Mintz
Center for Theoretical Studies
University of Miami

Laurence Mittag
Center for Theoretical Studies
University of Miami

Lars Onsager
Center for Theoretical Studies
University of Miami

Edwin E. Salpeter
Laboratory of Nuclear Studies
Cornell University

Julian Schwinger
Department of Physics
University of California at Los Angeles

George Soukup
Center for Theoretical Studies
University of Miami

Ichiji Tasaki
National Institute of Mental Health
Laboratory of Neurobiology

Edward Teller
Lawrence Berkeley Laboratory
University of California

Georges Ungar
Department of Anesthesiology and Pharmacology
Baylor College of Medicine

Jan Peter Wogart
Institute of Inter-American Studies of the Center for Advanced International Studies
University of Miami

OBSERVERS:

George Adelman
Managing Editor and Librarian
Neurosciences Research Program
Massachusetts Institute of Technology

Robert Lind
Department of Physics
Florida State University

F. David Peat
National Research Council of Canada

SUBJECT INDEX